HOW TO LIVE IN
SPACE

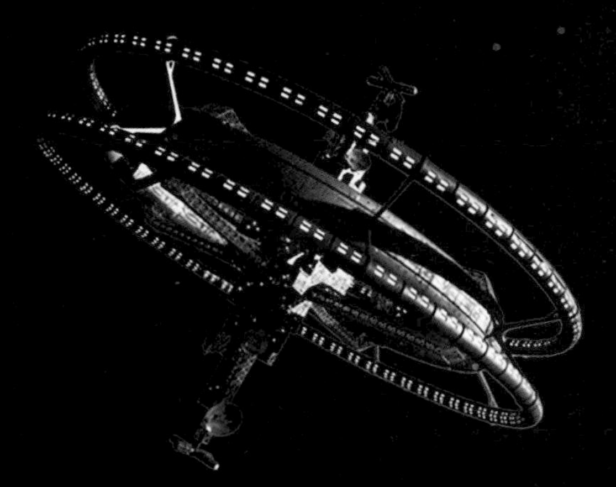

This book was designed and produced by
Carlton Publishing Group
20 Mortimer Street
London W1T 3JW

Published in North America by Smithsonian Books

This book may be purchased for educational, business, or sales promotional use. For information, please write: Special Markets Department, Smithsonian Books, P.O. Box 37012, MRC 513, Washington, DC 20013

Library of Congress Cataloging-in-Publication Data

Names: Stuart, Colin (Science writer), author.
Title: How to live in space : everything you need to know for the not-so-distant future / Colin Stuart.
Description: Washington, DC : Smithsonian Books, 2018. | Includes index.
Identifiers: LCCN 2018011524 | ISBN 9781588346384 (paperback)
Subjects: LCSH: Manned space flight--Popular works. | Interplanetary voyages--Popular works. | BISAC: SCIENCE / Applied Sciences. | TECHNOLOGY & ENGINEERING / Aeronautics & Astronautics.
Classification: LCC TL793 .S86 2018 | DDC 629.45--dc23 LC record available at https://lccn.loc.gov/2018011524

Manufactured in China, not at government expense

22 21 20 19 18 5 4 3 2 1

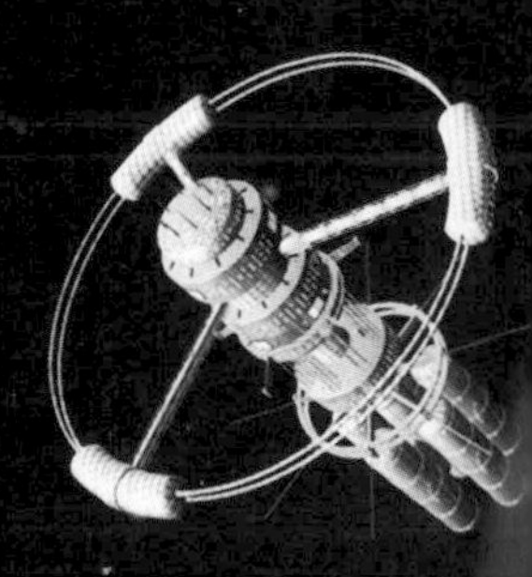

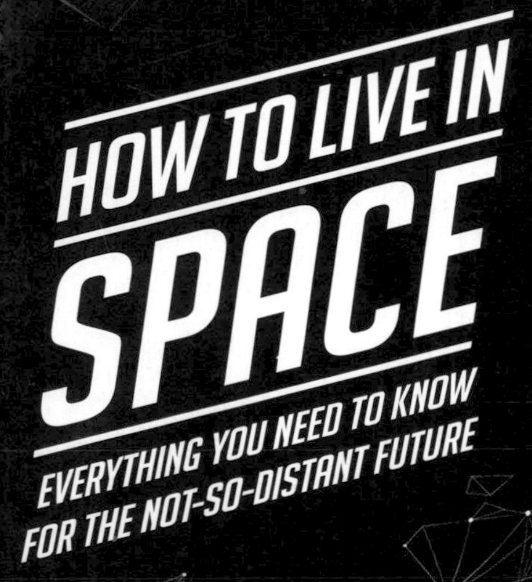

HOW TO LIVE IN SPACE

EVERYTHING YOU NEED TO KNOW FOR THE NOT-SO-DISTANT FUTURE

COLIN STUART

Smithsonian Books
Washington, DC

PART 1

TRAINING

PART 2

LIVING IN SPACE

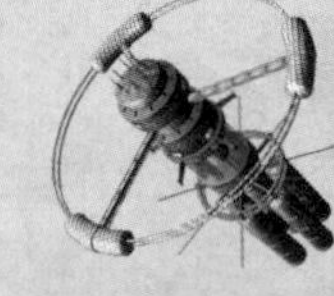

PART 3

THE FUTURE

OUR DESCENDANTS COULD LOOK BACK TO TUESDAY, FEBRUARY 6, 2018, AS THE DAY WHEN EVERYTHING CHANGED.

On a remarkable sunny afternoon in Florida, SpaceX's Falcon Heavy rocket surged into the sky and entered the record books as the most powerful launch vehicle in decades. Onlookers watched openmouthed as two of its boosters simultaneously returned to the ground in a carefully choreographed landing. The Falcon Heavy's payload was SpaceX founder Elon Musk's own red Tesla sports car, with a spacesuit-clad mannequin called Starman in the driver's seat, complete with David Bowie's "Space Oddity" playing on repeat. It was fired out of Earth orbit toward the asteroid belt. The next day, the newspapers' front pages around the world carried the now iconic image of a car floating in front of the blue marble of Earth.

What makes this event such a pivotal point in human history? After all, we've been launching objects into space since the 1950s. The difference is that this landmark feat was achieved by a private company, not a national government agency. Just as the launch of Sputnik 1 by the Soviet Union in 1957 fired the starting gun on a Space Race that eventually led to humans setting foot on the Moon, there is a real feeling in the space community that the Falcon Heavy launch is the trigger event for a revitalized second Space Race that will see humans walk on Mars later this century. Not only that, but ordinary people like you and me will get to go to space, as competing companies, such as Virgin Galactic and Blue Origin, drive down the price of access to orbit. The era of space tourism is here. This is the beginning of humans spreading out from their home planet and colonizing the Solar System. People will be talking about the events of that day for centuries to come.

Yet crossing a new frontier is never easy. For us humans, traveling into space creates its own unique set of challenges. Removing your body from the environment in which it has evolved to thrive can make everyday activities incredibly arduous. You'll have to face unprecedented dangers, but you'll also see unparalleled beauty. Consider this book your space travel manual. It contains all the knowledge you'll need to thrive in this new world order. What is it like to train to be an astronaut? How do you breathe in space? What does the lack of gravity do to your bones? How long will the journey to Mars take, and what will you do when you get there? All of these questions and more are answered in the pages that follow.

As a would-be space traveler, you'll probably encounter a wall of skeptics who question this new age. They argue that space exploration is a waste of money and that we have more pressing concerns to address right here on Earth. Such a view is more than a little myopic. The Apollo Moon landings sparked an interest in science and engineering for a generation. Those researchers now work in a diverse range of fields, solving wide-reaching problems well beyond the space sector. In the decades to come, a new wave of problem solvers will cite the moment they saw a cherry-

red sports car floating above the planet as the moment they were inspired to reach for the stars.

The universe out there is so different from our lives down here that we have to throw away the rule book. Space olts us into thinking in an entirely different way. By asking he biggest questions possible, and working creatively to answer them, we get inventive. Solutions to space-related problems that at the time seemed otherworldly are now routinely being used to benefit us all. Telescope software is being used to diagnose cancer. A robotic arm designed for he International Space Station (ISS) has been adapted for surgery. Mars rover technology can now be found in gas-eak detectors and insulin pumps for people with diabetes. Space research doesn't take away from the rest of life—it adds to it. By reaching out into space, we not only ensure hat we're constantly breaking new ground, but a permanent presence beyond Earth will give the human race the ultimate nsurance policy if the worst happens to our original planet.

I am so excited about what the decades ahead will bring, and I hope you enjoy joining me on the ultimate human dventure. May we never stop reaching for the stars.

∧ Two of the boosters from the Falcon Heavy rocket landing simultaneously back on Earth.

↖ Elon Musk's red Tesla sports car with Starman in the driver's seat and Earth behind.

∧ Canadarm on a Space Shuttle—the technology has been adapted for use in surgery.

INTRODUCTION

WHERE EXACTLY IS SPACE?

AT FIRST GLANCE THE ANSWER MIGHT SEEM SIMPLE: "UP THERE." BUT WHERE EXACTLY DOES EARTH END AND SPACE BEGIN? YOU MIGHT BE SURPRISED TO HEAR THAT SPACE STARTS JUST 60 MILES [100 KM] ABOVE YOUR HEAD.

That's the location of the Kármán line, named after Hungarian-American engineer Theodore von Kármán (1881–1963). At this altitude, our planet's atmosphere is so thin that conventional aircraft cannot generate lift on their wings unless they travel at a speed that also places them in orbit around Earth. The Fédération Aéronautique Internationale (FAI) recognizes it as the point where aeronautics become astronautics. Anyone traveling beyond it is as an astronaut. More than five hundred people have made that journey since Yuri Gagarin became the first person in space in 1961 (see box, below). In reality, you have to get above 100 miles (160 km), or atmospheric drag will cause you to fall from orbit quickly.

YURI GAGARIN AND VOSTOK 1

On April 12, 1961, Soviet fighter pilot Yuri Gagarin (1934–68) made history when he became the first human to cross the Kármán line and orbit around Earth. Launching from the Baikonur Cosmodrome in the steppes of Kazakhstan, his whole *Vostok 1* mission lasted just 108 minutes, including his landmark 92-minute lap of the planet.

However, it nearly ended in disaster. Vostok 1's retrorockets fired to begin its descent while over the West African nation of Angola. Shortly afterward, the two halves of the craft failed to separate correctly and began to gyrate wildly. Fortunately, the vibrations weren't enough to jeopardize the mission, but they did lead to Gagarin touching down 175 miles (280 km) west of the intended landing site.

Gagarin would later recall how he landed close to a farmer and his daughter, who were initially shocked to see an alien-looking figure in an orange space suit and white helmet who had fallen from the sky. He reportedly told them: "Don't be afraid, I am a Soviet like you who has descended from space and I must find a telephone to call Moscow!"

He became a worldwide celebrity, but died in 1968 in a fighter jet crash at the age of thirty-four.

^ In 1961, Soviet cosmonaut Yuri Gagarin became the first human to orbit Earth.

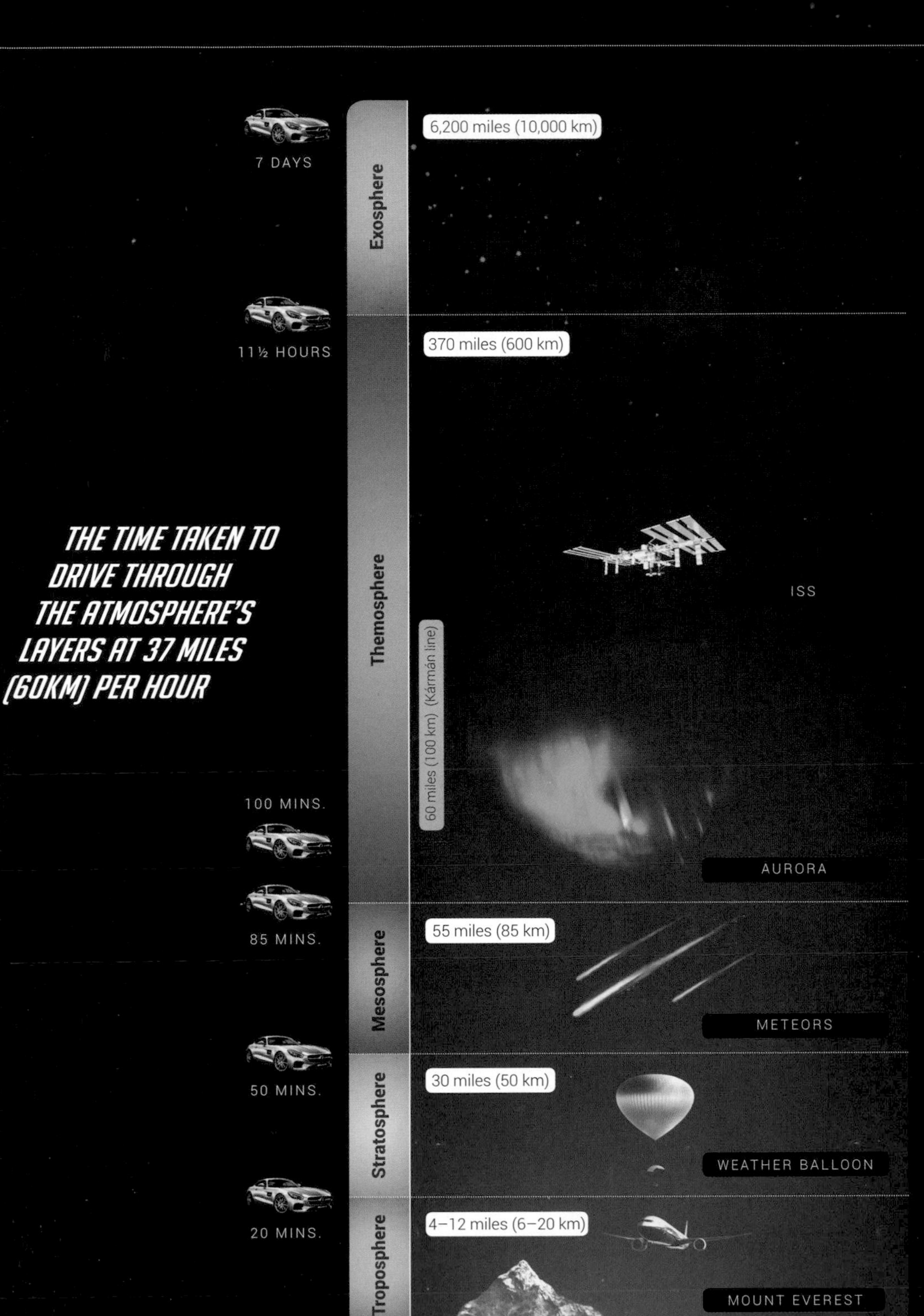
THE TIME TAKEN TO DRIVE THROUGH THE ATMOSPHERE'S LAYERS AT 37 MILES (60KM) PER HOUR
7 DAYS
11½ HOURS
100 MINS.
85 MINS.
50 MINS.
20 MINS.
Exosphere
Themosphere
Mesosphere
Stratosphere
Troposphere
6,200 miles (10,000 km)
370 miles (600 km)
60 miles (100 km) (Kármán line)
55 miles (85 km)
30 miles (50 km)
4–12 miles (6–20 km)
ISS
AURORA
METEORS
WEATHER BALLOON
MOUNT EVEREST

It is important, however, to realize that the Kármán line doesn't mark the edge of Earth's atmosphere. In fact, it sits partway through a layer of the outer atmosphere called the thermosphere, which extends between 55 and 370 miles (90 and 600 km) above sea level. Above that is the exosphere, which stretches out to 6,200 miles (10,000 km) from Earth's surface. If the edge of the exosphere marked the point where space begins, the ISS (approximately 250 miles/400 km up) and many of our communications and weather satellites wouldn't be in space!

The region in which satellites orbit Earth is split into four bands based on altitude. Objects up to 1,250 miles (2,000 km) above the ground are said to be in low Earth orbit (LEO). It typically takes 92 minutes to orbit Earth from this height. The 24 astronauts who were part of NASA's Apollo program to the Moon are the only humans to ever leave LEO. You'll also find the famous Hubble Space Telescope here.

Next is medium Earth orbit (MEO), extending from the top of LEO to just over 21,750 miles (35,000 km) above sea level. The most famous residents of this locale are the satellites of the Global Positioning System (GPS) that orbit the planet every twelve hours or so. Then you'd come to the geosynchronous and geostationary satellites, which supply us with satellite television, among other things. Sitting at the special altitude of 22,236 miles (35,786 km), they orbit in exactly the same time it takes for Earth to spin, ensuring that they stay visible in our sky at all times. Finally, any object above 22,235 miles (35,786 km) is said to be in high Earth orbit (HEO). Satellites in this region are rare but can include astronomical observatories or nuclear weapons detectors trained on the ground.

*Up to end of 2017

Note: The data shows each time a person crossed the Kármán line, so a single astronaut can appear more than once.

DECADE	MALE	FEMALE
1950s	0	0
1960s	69	1
1970s	82	0
1980s	217	15
1990s	400	69
2000s	277	40
2010s*	161	21

AURORAE AND SHOOTING STARS

> The beautiful spectacle of Aurora Borealis (northern lights) dancing across the Kármán line.

Aurorae are one of greatest spectacles of nature—polar displays known as the Northern lights (Aurora Borealis) and Southern lights (Aurora Australis). These dancing curtains of intense light pirouette across the Kármán line, occurring 55 to 80 miles (90–130 km) above our heads. They are caused by electrical currents flowing through the base of the thermosphere, generated by interactions between material ejected by the Sun and Earth's magnetic field. Oxygen atoms in the atmosphere jettison the extra energy they receive from the currents in the form of lustrous green light. Vivid displays are sometimes accompanied by audible sounds, including pops and muffled bangs.

Meteors—colloquially known as "shooting stars"—are another standout atmospheric phenomenon; however, they occur at a slightly lower altitude, in the mesosphere. Here, the atmosphere is thick enough that friction causes cosmic dust grains falling to Earth to incinerate as they streak across the sky in short, brilliant bursts. Meteors frequently occur in annual showers when our planet passes through clouds of debris strewn across the Solar System by meandering comets. Perhaps the most famous shower is the Perseids, which occurs each August as the result of dust from the comet Swift-Tuttle.

INTRODUCTION

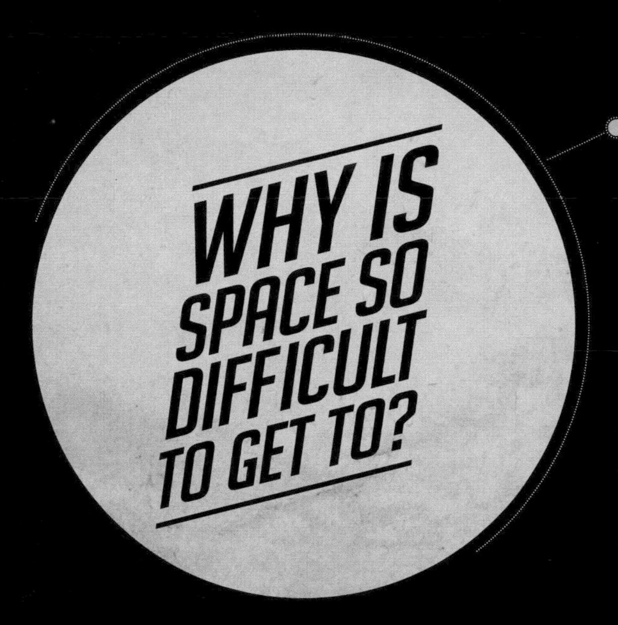

IF SPACE IS ONLY 60 MILES (100 KM) AWAY—A DISTANCE HUMAN BEINGS TRAVEL ALL THE TIME IN THEIR DAILY LIVES—WHY HAVE ONLY A TINY PROPORTION OF THE WORLD'S POPULATION MADE THE JOURNEY THERE?

The simple answer can be given in one word: gravity. Earth is a heavy planet, tipping the scales at 13.17 septillion pounds (5.9 septillion kg)—a septillion is a number with twenty-four zeros. Since the days of Isaac Newton (1643–1727), we've known that any two objects with mass are attracted to each other by the force of gravity. The strength of that force depends on the masses of the two objects and the distance between them. To get into space, you need to overpower Earth's mighty gravity and climb into the sky faster than it can pull you back down.

KONSTANTIN TSIOLKOVSKY

Born in Russia in 1857, Tsiolkovsky is widely credited as the father of space flight. His work, including more than ninety published papers, was influential in making space travel a reality. He devised the rocket equation that calculates the change in speed of a rocket from the velocity of material leaving its exhaust and the mass of the rocket before and after launch.

Decades ahead of his time, he also studied many of the features we'll come across in the rest of this book, including multistage rockets, air locks, and space stations. He even considered the possibility of space elevators, inspired by the construction of the Eiffel Tower in 1887. He reportedly said "Earth is the cradle of humanity, but one cannot remain in the cradle forever." A crater on the Moon is named after him in recognition of his landmark work.

Despite his incredible insights, Tsiolkovsky never actually made a rocket. He spent a lot of time living a hermit's existence in a log house on the outskirts of a town 125 miles (200 km) from Moscow. At the age of 10, he'd contracted scarlet fever, and it left his hearing severely impaired. He died in 1935 at the age of 78.

> Konstanin Tsiolkovsky was the founding father of rocketry and worked on the math behind launching rockets.

The speed required to achieve this is known as Earth's escape velocity, and it is a substantial 7 miles (11.2 km) per second. Technically, you can get into space traveling as slow as 1,000 yards (1 km) per second, but you'd have to put in increasingly more energy to maintain that speed as Earth's gravity acts to slow you. If you launch at escape velocity instead, you'll reach space before you're halted. It doesn't depend on the mass of the object launched—it is the same for a golf ball or the Saturn V rocket that took the Apollo astronauts to the Moon. But the heavier your payload, the more fuel you'll need to accelerate your rocket to escape velocity.

It's an expensive endeavor. Traditionally, it has cost around $20,000 for every kilogram—about 2¼ pounds—you want to get into low Earth orbit (LEO). Much of that cost is because rockets haven't been reusable in the past. Each time you launch something into orbit, you have to pay for a new rocket. However, new companies, such as Elon Musk's SpaceX, are revolutionizing the rocket industry and driving down the price. Their Falcon 9 rocket can deliver a payload to LEO and land back on a barge floating at sea to be collected and reused. Without the substantial overheads of new rockets, SpaceX can get a kilogram to LEO for less than $3,000—a huge reduction.

NASA has already used SpaceX rockets to deliver cargo to the ISS, and astronauts will probably be next. That means, in the decades to come, space will no longer be the preserve of a lucky few. SpaceX is part of a growing legion of companies knocking down the barriers to space tourism (see pp. 124–27.). LEO will soon become a playground for the ordinary person as access to space opens up and it becomes a much easier place to reach.

> Illustration drawn by Henri de Montaut in *From the Earth to the Moon,* by Jules Verne.

Long before we could actually reach space, we spent centuries dreaming of traveling there. These days, rockets are the transportation of choice, but over the years we've contemplated some pretty wacky ways of leaving Earth behind. As far back as the second century CE, Syrian author Lucian of Samosata was writing about people fired to the Moon by a water spout. In the 1516 story *Orlando Furioso*, the title character goes to the Moon in a flaming chariot, whereas birds called gansas pull a person there in Francis Godwin's 1638 work *The Man in the Moone*.

The invention of the hot air balloon in the late 1700s saw airships dominate our space travel ideas for a century, before science fiction writers such as Jules Verne turned to a space cannon catapulting people into orbit in stories such as *From the Earth to the Moon* (1865). Verne's work, in particular, fired the imaginations of early rocketeers, such as Konstantin Tsiolkovsky (see p.13), Hermann Oberth, Wernher von Braun (see p.46), and Robert Goddard (see p.46). Tsiolkovsky calculated that the accelerations required by Verne's cannon would kill any astronauts aboard, so he turned his attention to the physics of rockets instead and jump-started the era of being able to get to space rather than just fantasizing about it.

- ASTRONAUT SELECTION
- UNDERWATER TRAINING
- CENTRIFUGE TRAINING
- THE VOMIT COMET
- FLIGHT TRAINING
- THE LAUNCH
- MISSION CONTROL
- ROCKET SEPARATION
- SPACE SICKNESS
- RENDEZVOUS AND DOCKING

ASTRONAUT SELECTION

IT IS THE CHILDHOOD DREAM OF MILLIONS, BUT DO YOU HAVE WHAT TAKES TO BECOME AN ASTRONAUT? SO FEW PEOPLE ACTUALLY GET TO GO TO SPACE THAT COMPETITION IS UNDERSTANDABLY FIERCE.

NASA ran an astronaut recruitment drive in the winter of 2015 to 2016. During that period, they attracted more than 18,000 applications—a record.

To get over the first hurdle, you need to meet some basic requirements. You must have a degree in a relevant area of science, engineering, mathematics, or medicine, along with at least three years of field experience. For pilots, that can be substituted for one thousand hours of pilot-in-command time in a jet aircraft. Teachers can also apply if they teach one of those scientific disciplines. You also need to meet exacting physical criteria (see box, right) and be a US citizen (dual nationality is acceptable). A panel of judges, including former astronauts, assesses those left after sifting out the applicants with inadequate educational backgrounds. A pool is then created containing the 500 best candidates. The judges are looking for all-rounders and team players, attributes that are often reflected in their choice of hobbies and interests as much as their work lives.

The next stage sees 120 lucky applicants invited for a first interview, but only half of them will be asked back for a second. Those sixty people undergo a grueling, week-long schedule of physical tests and medical examinations. Only eight to fourteen candidates will move on to start several years of astronaut training (see pp. 20–35). So, of the thousands that applied, just 0.07 percent succeed.

To add to the challenge, astronaut recruitment drives are rare. Since the European Space Agency (ESA) was founded

∧ NASA astronaut Jim Lovell inside his Gemini 12 spacecraft in 1966.

Joshua Kutryk and Jennifer Sidey, members of the Canadian Astronaut Corps.

The six astronauts selected by the European Space Agency in 2009.

in 1975, there have been only three intakes: 1978–79, 1991–92, and 2008–09. In the last round, ESA received 8,413 applications that met their eligibility criteria. A thousand underwent psychotechnical tests, and just six astronauts emerged from the process.

So how much do successful candidates earn? Maybe not as much as you'd expect for a job where you're placing yourself in a considerable amount of danger. While space agencies don't directly publish astronauts' salaries, the relevant pay bands stretched from $64,724 up to $141,715 in 2012. Former International Space Station (ISS) commander Chris Hadfield confirmed as much when he revealed that an astronaut can earn up to $150,000, depending on his or her experience. In Europe, trainee astronauts enter at the A2 pay grade, worth almost 54,000 Euros ($57,000). ESA astronauts are often promoted to the A4 grade after their first space flight, worth 77,000 Euros ($81,000) a year. There is more money to be had from appearances and book deals, but few do it for the wealth. Being an astronaut is reward enough in itself.

ASTRONAUT PHYSICALS

It goes without saying that you need to be physically fit to become an astronaut. NASA's long-duration astronaut physical dictates that you must meet the following criteria:

- Your blood pressure cannot exceed 140/90 in a sitting position
- Your eyesight needs to be correctable to 20:20
- You must be between 5 feet 2 inches to 6 feet 3 inches (1.57 and 1.91 m) in height
- You must be able to swim 246 feet (75 m) without stopping and then swim the same distance in a flight suit and tennis shoes, with no time limit
- Still wearing the suit and shoes, you must be able to tread water for 10 minutes without stopping.

The European Space Agency (ESA) has similarly strict guidelines. Applicants must:

- Be free from any disease and dependency on drugs, alcohol, or tobacco
- Have the normal range of motion and functionality in all joints and visual acuity in both eyes of 100 per cent (20/20) either uncorrected or corrected with lenses or contact lenses
- Be free from any psychiatric disorders
- Demonstrate cognitive, mental, and personality capabilities to enable them to work efficiently in a highly demanding intellectual and social environment

Life-size mock-ups of the International Space Station modules are used for astronaut training.

JIM LOVELL AND THE LUMINOUS ALGAE

v Like many early astronauts, Jim Lovell (b. 1928) was a fighter pilot first.

In the early days of space flight, all astronauts were selected from an adventurous, reliable band of military pilots. Jim Lovell, who would later command the ill-fated Apollo 13 lunar mission (see p. 59), first dreamed of rockets and space as a high-school student after reading the stories of Jules Verne. He joined the navy instead as a pilot, before eventually realizing his astronautical ambitions.

In 1953, still learning the ropes as part of the Composite Squadron Three aircraft carrier group near Japan, he attempted to land his F2H Banshee on the deck of the USS Shangri-La. His instrument panel, however, short-circuited, plunging him into darkness. The quick-thinking Lovell spotted fluorescent algae in the ocean that were being churned up by the ship's wake. The tiny, glowing organisms guided him all the way in. That kind of instinctive ingenuity would see him selected as a test pilot and later chosen for the second intake of astronauts picked by NASA to form the crew of the Gemini program. Over an eleven-year astronaut career, Lovell spent more than seven hundred hours in space, becoming the first person to fly around the Moon twice: once with Apollo 8 and again on Apollo 13.

UNDERWATER TRAINING

LOCATED 6 MILES (10 KM) OFF THE COAST OF FLORIDA, NESTLED 62 FEET (19 M) DOWN ON THE SEABED OF THE ATLANTIC OCEAN, SITS ONE OF THE MOST IMPORTANT ASTRONAUT TRAINING FACILITIES ON EARTH.

Here, you'll find the Aquarius Reef Base, which is operated by Florida International University. Initially designed as underwater accommodation for biologists studying the local coral reef, it has also become a home away from home for astronauts to hone their skills.

The sea has claimed Aquarius as its own since it was installed in 1986—its outer shell is encrusted with an impressive array of marine life. About the size of a school bus, with a kitchen, laboratory, and bunk beds for six, Aquarius is made up of three separate compartments comprising an entry lock situated between the wet lock and the reinforced main compartment, in which surface atmospheric pressure is maintained. The base is equipped with wi-fi, and portholes let residents look out at the wonderful array of local sea life, including groupers, sharks, and sponges.

International crews of astronauts, engineers, and scientists typically spend ten days to two weeks on board as part of NASA Extreme Environment Mission Operations (NEEMO) missions. The crew members spend up to nine hours a day outside the habitat, often diving to depths of nearly 100 feet (30 m).

The buoyancy experienced underwater by these "aquanauts" is a great way to simulate the weightlessness of outer space without anywhere near the level of cost or risk. Different scenarios, from Mars's gravity to that found on an asteroid, can be simulated by varying the

∧ Astronauts train for spacewalks underwater to simulate the effects of weightlessness.

↗ An international crew of astronauts and technicians outside the Aquarius Habitat.

Astronauts training for extravehicular activities (EVAs)—otherwise known as spacewalks—are often submerged in a huge swimming pool. Located at NASA's Johnson Space Center in Houston, Texas, this Neutral Buoyancy Laboratory (NBL) is 202 feet (61.5 m) long, 102 feet (31 m) wide, and 40 feet 6 inches (12.3 m) deep. It can hold 6.2 million gallons (23.5 million liters) of water. Containing a model of the International Space Station, it is overlooked by command rooms where mission controllers can monitor the astronauts' activity and communicate with them directly.

Astronauts have racked up more than one-hundred thousand hours in the NBL since it was established in 1995. There's never been an accident, partly thanks to the safety divers, who are also constantly monitoring the astronauts. Trainees wear special suits, similar to their real spacesuits, tailored to their bodies, based on 36 measurements of their frames and 46 measurements of their hands. The suits are connected to two tethers, each 85 feet (26 m) long, supplying them with air to breathe. It contains 46 percent oxygen (compared to 21 percent in normal air) to help reduce the risk of decompression sickness. Shorter tethers keep them in place along handrails while they practice intricate tasks, including repairing the outer hull of the ISS.

∧ NASA astronaut Reid Wiseman in the Neutral Buoyancy Laboratory (NBL) back in 2012.

weight of equipment strapped to the divers. Many missions test and evaluate tools and techniques for future spacewalks. In 2015, they tried out technology designed by ESA that showed astronauts the next step in a complicated series of tasks on a head-mounted display. It meant the astronauts didn't have to take their eyes away from the intricate task on hand. Often crew members have walked in space for real, which helps to bring home the realism of these exercises for the new trainees. Time delays are introduced into communications to simulate being a considerable distance from Earth, particularly when mocking up activities on distant asteroids or in Mars orbit.

However, a subaqua existence isn't without its dangers. A diver died in 2009—an exercise had compromised his breathing apparatus. The Aquarius Reef Base has been damaged by hurricanes on several occasions, too. Hurricane Gordon in 1994 caused one of its generators to catch fire, and the crew had to ascend a rescue line in 15-foot (4.6-m) seas. Its anchors were broken by Hurricane Rita in 2005. Then, in 2017, Hurricane Irma blew Aquarius's life support moorings 14 miles (22.5 km) away.

> NASA astronaut Serena Aunon moving tools and equipment underwater as part of NEEMO 20.

SEA SURVIVAL TRAINING

As an astronaut, you have to be prepared for every eventuality. This includes a landing at sea. While the Soyuz capsule is intended to touch down on land, any number of difficulties during reentry could see you splash down in the ocean instead. After all, most of Earth's surface is water. In 1976, the Soyuz 23 capsule came down off target and into a freezing saltwater lake. In contrast, the NASA Apollo missions landed in the sea by design.

All astronauts undergo some level of sea survival training. Often, a mock-up capsule is winched into water and trainees clamber inside. The ideal scenario is that your capsule remains watertight and you can wait it out for a few days while the search-and-rescue party find you. But a damaged landing module could take on water and start sinking, just as Gus Grissom's Mercury capsule did in 1961. So astronauts are schooled in emergency escapes.

If they have more time, astronauts are taught to change from their flight suits into rubber replacements designed to keep their body temperature from dropping too low and aid buoyancy. They then move into inflatable rafts and try to call for help on satellite telephones.

NASA

NASA'S AMES RESEARCH CENTER IN SILICON VALLEY, CALIFORNIA, IS HOME TO THE 20-G CENTRIFUGE—A CRUCIAL TOOL IN TRAINING ASTRONAUTS TO COPE WITH THE RIGORS OF SPACE TRAVEL.

The centrifuge prepares astronauts' bodies for the punishing G-forces (see box, right) that they will experience during launch and reentry of Earth's atmosphere.

The acceleration experienced by astronauts during launch makes them feel heavier than they are. They're pressed into their seat, much like you are when you accelerate in a car. A typical lift-off will see astronauts experience G-forces of 3 g—three times the normal acceleration toward the ground due to gravity (known as 1 g). Reentry is even more extreme, with astronauts often pulling 8 g for about thirty seconds. Sometimes it can be more, or for longer, just as when the three astronauts of the Soyuz TMA-11 mission returned from the International Space Station in 2008. They experienced 8 g for sixty seconds during a ballistic reentry—a steeper entry than normal due to a malfunction.

The Ames centrifuge recreates these accelerations by rotating a 58-foot (17.7-m) arm up to 50 times a minute around a circular room. Renovated in the early 1990s, it has a mass of 1,200 pounds (544 kg) and contains three compartments known as "cabs." Rapid accelerations are felt at the ends as the arm accelerates around the track. It can create G-forces of 12.5 g with human occupants, or up to 20 g without. Other space agencies have their own centrifuges, including one housed at Star City outside Moscow.

G-FORCES, MASS, AND WEIGHT

Although people often interchange the terms "mass" and "weight" in everyday conversation, the terms have different meanings in physics. Mass is a measure of how much stuff you are made of—the total number of pounds or kilograms all your atoms add up to. You might say you're losing weight because you're on a diet, but in fact you're losing mass. Your mass doesn't change by getting on a rocket or sitting in a centrifuge. Weight, on the other hand, is how heavy you feel, and that depends on the acceleration (or deceleration) you are experiencing. Your weight is your mass multiplied by that acceleration and is measured in newtons (N).

During our normal day-to-day activities, we will feel an acceleration toward the ground due to the gravity of Earth. This is known as 1 g. However, we've all probably experienced greater accelerations. Speed up in a car and you feel pressed down in your seat—you feel heavier. That's because an acceleration greater than 1 g makes you weigh more. The highest G-force you can experience on a rollercoaster is on the Tower of Terror ride at Gold Reef City in South Africa—an impressive 6.3 g.

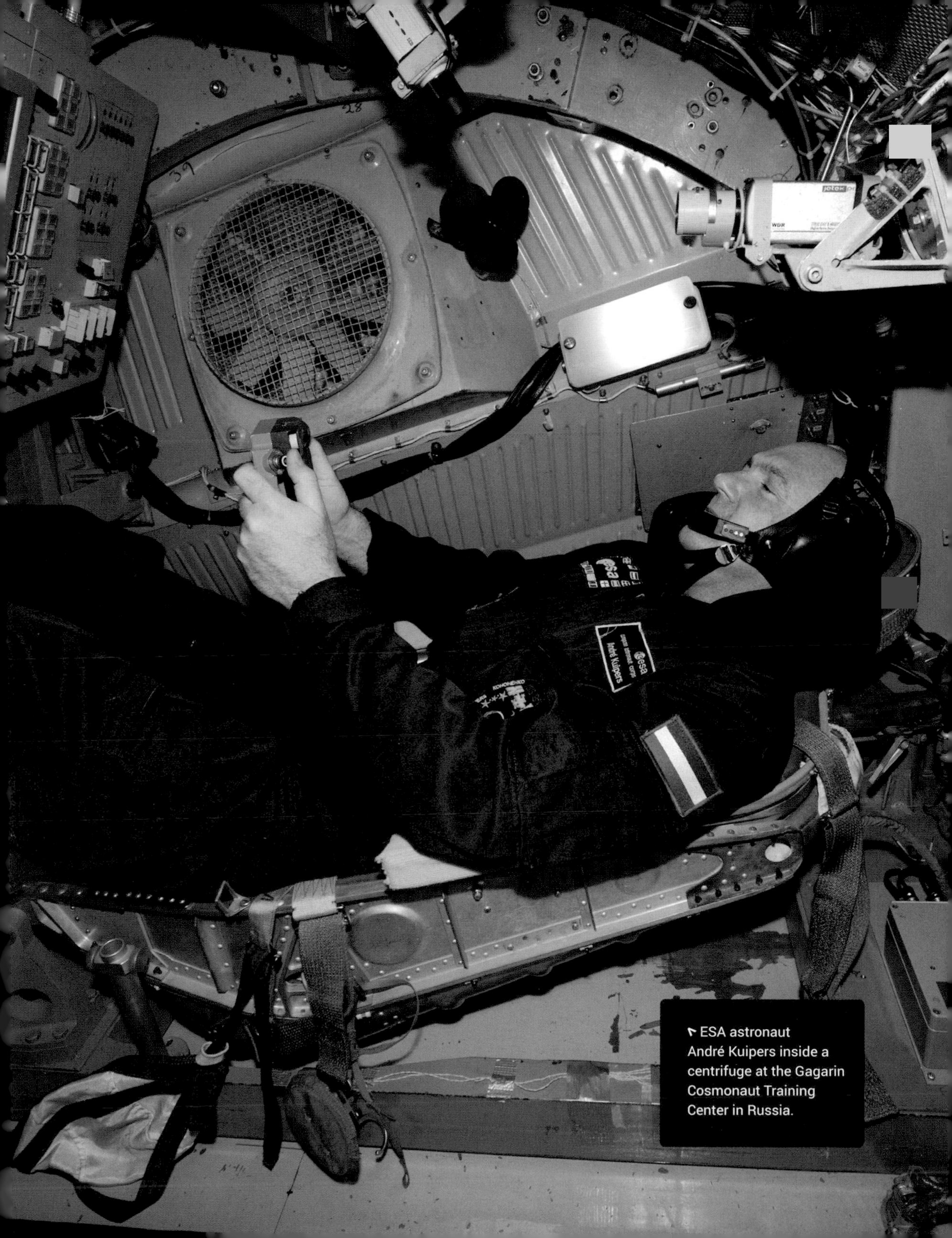

ESA astronaut André Kuipers inside a centrifuge at the Gagarin Cosmonaut Training Center in Russia.

Few astronauts are fans of centrifuge training. John Glenn—the first US astronaut to orbit Earth—called it "dreaded" and "sadistic." Apollo 11 astronaut Michael Collins wasn't enamored either, calling it "diabolical."

The need for protection against G-forces first became clear as early as World War I, when pilots flying fighter planes would faint due to the G-forces that they experienced during the twists and turns of dogfights. Extreme G-forces can alter the distribution of blood around the human body. The last thing you want is for all your blood to pool in your legs and starve your head of oxygen. Cut off the supply to your eyes and your vision begins to suffer—an effect known as "graying out." Losing consciousness—"blacking out"—is a real possibility, too. This problem is partly mitigated by orientating the astronauts within the spacecraft so that the G-force is felt through their chest instead of head to toe. Astronauts also wear G-suits, which apply pressure to key areas to help keep their blood moving.

Much of an astronaut's time in a centrifuge is spent learning about the muscle flexing and breathing techniques that can complement their G-suits. Breathing out is fine, but refilling your lungs is next to impossible under the crushing weight. Instead, astronauts learn to keep their lungs almost completely full, restricting themselves to short, sharp inhales and exhales.

∧ The 20 g centrifuge at NASA's Ames Research Center in California.

v Riders on roller coasters often experience G-forces many times their own weight.

JOHN STAPP (1910–1999)

> The effect of G-forces on John Stapp's face can be seen in this still from one of his experiments.

The remarkable work of US Air Force physician John Stapp led to some of the earliest insights into the effects of high G-forces on the human body. Long before modern centrifuges, Stapp built a rocket-power sled at Edwards Air Force Base in Southern California nicknamed the "Gee Whiz." Strapped to the sled, he traveled along the track at exceptionally high speeds before decelerating quickly to a halt.

Along the way he cracked his ribs, broke his wrist twice, and lost dental fillings, but he survived all his runs, eventually withstanding record-breaking (albeit short-lived) forces of 46 g. During that final landmark test in 1954, he decelerated from 632 miles (1,017 km) per hour—90 percent of the speed of sound—coming to a stop in a little over a second. His body momentarily weighed the equivalent of nearly 3.8 tons (3,500 kg). It saw him break the land speed record and secured him a place on the front cover of *Time* magazine, billed as "The Fastest Man on Earth."

Most important, it improved safety on board fighter jets, allowing for humans to push the envelope of what was possible and eventually leading to some of those fighter pilots becoming the first generation of astronauts.

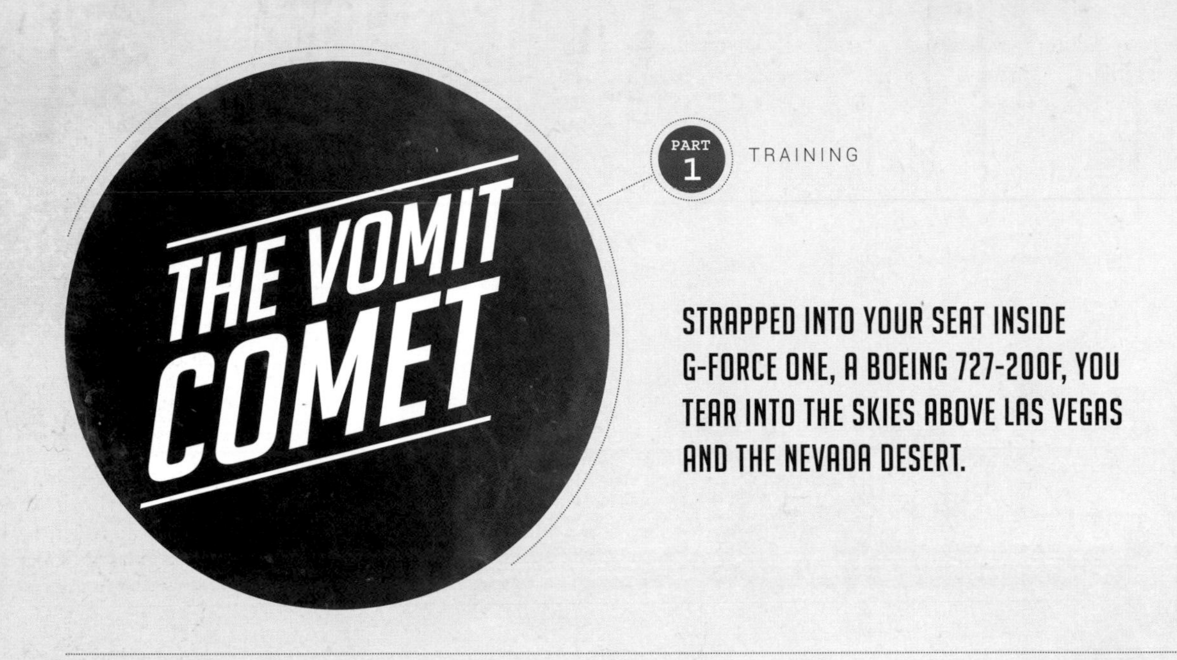

THE VOMIT COMET

STRAPPED INTO YOUR SEAT INSIDE G-FORCE ONE, A BOEING 727-200F, YOU TEAR INTO THE SKIES ABOVE LAS VEGAS AND THE NEVADA DESERT.

The next two or three hours are going to be unlike anything you've ever experienced. You're a passenger on the fabled Vomit Comet—a hollowed-out airliner set to perform a series of parabolic arcs in order to re-create the weightless environment of space (see box, p.30).

Soon you'll be floating, flipping, and flying just like an astronaut. You just have to wait thirty minutes to reach the right altitude.

When at 24,000 feet (7,300 m), the pilot pulls the plane up sharply at a 47-degree angle, climbing to 32,000 feet (9,750 m). During the sharp ascent, you'll experience G-forces of 1.8 g, making you feel almost twice as heavy as normal. As the plane begins to level out at the top of the climb, you and your fellow passengers will experience twenty to thirty seconds of weightlessness (0 g) before the pilot pulls out of the resulting dive on the way back to 24,000

PARABOLIC FLIGHT

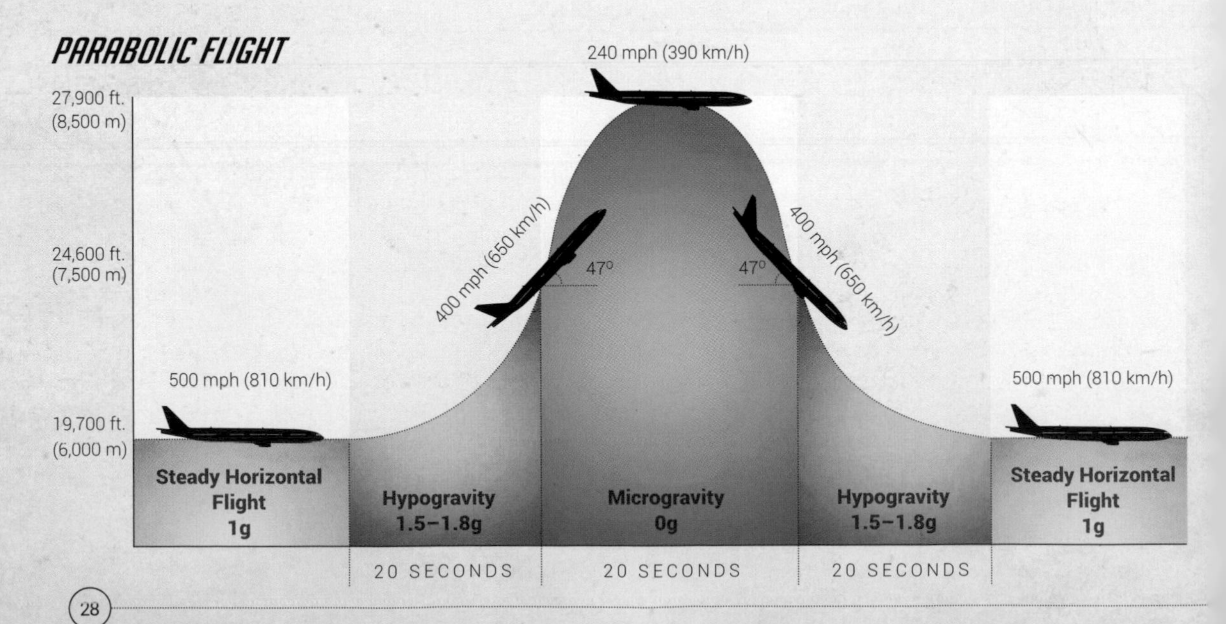

feet again (see diagram). This pattern is repeated at least fifteen times so you can get more proficient with practice. Constant updates come from the crew, with countdowns to the next phase of flight.

Apparently, at first, novices instinctively try to swim. Once passengers adapt, they move on to throwing balls to each other, eating pieces of floating food or maneuvering themselves around the cabin using the handrails on the padded walls. However, there is a downside: The Vomit Comet often makes passengers drastically unwell—hence the frank nickname.

The experience used to be reserved for trainee astronauts, but the company that flies would-be NASA space travelers—Zero G Corp—also offers commercial flights for about $5,000. Other space agencies have similar aircraft. The European Space Agency (ESA) uses a modified Airbus A310, the Canadian Space Agency a Falcon 20. In 2016, former ESA astronaut Jean-François Clervoy flew on the A310, recreating lunar gravity while wearing a virtual reality headset to farther enhance the feeling of being on the Moon.

In the United States, the Vomit Comet has a long history, predating even NASA (which was formed in July 1959). From 1957 to 1967 the US Air Force flew various planes on parabolic arcs to simulate weightlessness. It was used to train the astronauts of the Mercury and Gemini missions, along with performing other nonhuman experiments in a weightless environment without the cost and risk of going into space. The KC-135A was a particular stalwart, in service between 1994 and 2004. NASA estimates that it carried more than 2,000 students and even made an appearance in the Hollywood blockbuster *Apollo 13* (see box, below).

∧ European Space Agency astronauts training on the Zero-G Airbus 300 in 2010.

< The parabolic arc flown by the Vomit Comet creates twenty to thirty seconds of weightlessness.

THE VOMIT COMET IN HOLLYWOOD

Moviemakers have long used Vomit Comets to create dazzling special effects in blockbuster movies.

In the 2017 movie *The Mummy*, the characters played by Tom Cruise and Annabelle Wallis are caught in the middle of a plane crash and scrabble around for parachutes as their aircraft hurtles toward the ground. While the scene may look like CGI, it was filmed on a specially constructed set inside the European A310 plane as it carried out successive parabolic dives. Cruise later told Jimmy Fallon on *The Tonight Show* that the crew were regularly throwing up between takes.

Directors also turn to parabolic flights when shooting scenes where astronauts are floating in space. In the 1995 movie *Apollo 13*, actors Tom Hanks, Bill Paxton, Kevin Bacon, and Gary Sinise took to the skies with director Ron Howard. Despite being prescribed antinausea medication, both Bacon and Sinise were ill. Paxton apparently took to weightlessness like a duck to water.

Filming on a Vomit Comet has its unique limitations. Each parabolic dive lasts no more than thirty seconds, so a lot of short scenes have to be stitched together to form a longer sequence if, like in *Apollo 13*, large parts of the movie show astronauts in a weightless environment.

All objects within a gravitational field fall at the same rate irrespective of their mass. Galileo said this in the seventeenth century, and Apollo 15 astronaut Dave Scott proved it in 1971, when a feather and a hammer he dropped hit the Moon's surface at the same time.

Imagine severing the cable of an elevator car. The car would plummet toward the ground, only stopping when it hits the floor. Now imagine someone pushed you down the elevator shaft. You'd fall at the same rate as the car just did. So, if you're inside the car when its cable breaks, you'd appear to float because both you and the car would free fall at the same rate.

Astronauts are weightless because they're falling around Earth at the same rate as the spaceship they are in, just like you and the elevator car. Except they're high enough never to hit the ground.

So how does an airplane create weightlessness? The Vomit Comet does it by briefly entering freefall at the top of a parabolic trajectory, leading to its passengers being in freefall, too.

v The late physicist Stephen Hawking during a flight on a Vomit Comet in 2007.

↘ The Airbus A300 zero-gravity plane during the upward phase of a parabolic arc.

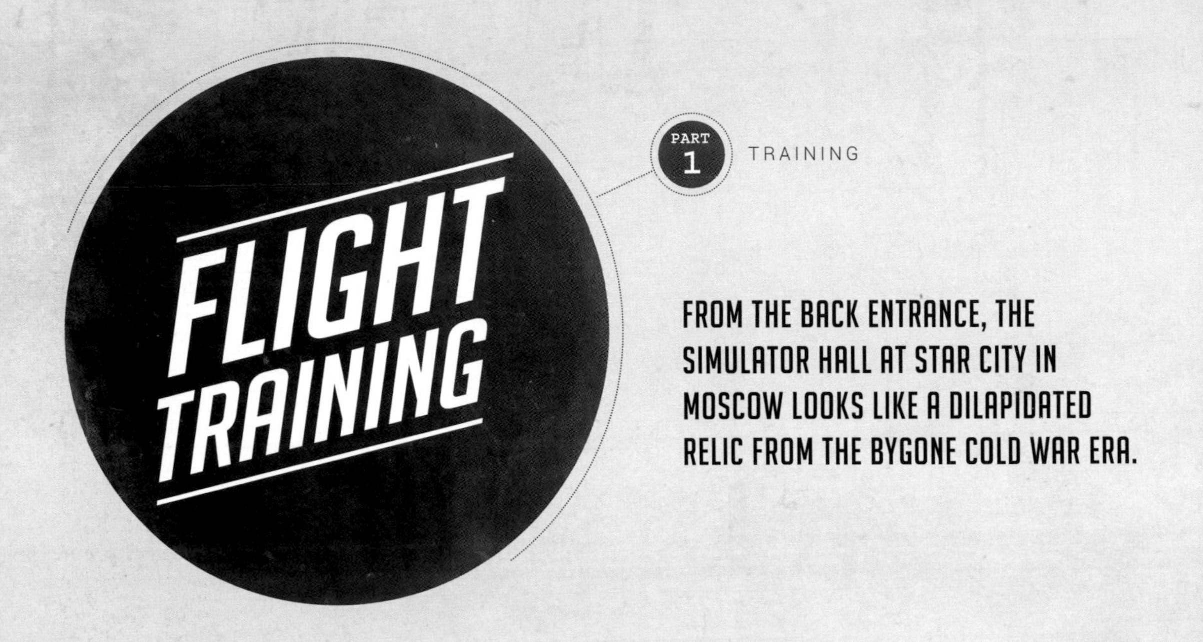

FLIGHT TRAINING

FROM THE BACK ENTRANCE, THE SIMULATOR HALL AT STAR CITY IN MOSCOW LOOKS LIKE A DILAPIDATED RELIC FROM THE BYGONE COLD WAR ERA.

With a grey facade and metal bars over the windows, it hardly seems like the home of one of the most advanced space simulators in the world. Yet inside is a highly realistic mock-up of the descent module of the mighty Soyuz spacecraft. Astronauts undergo full training inside, complete with flight suits and communication caps that let them communicate with Mission Control.

The whole exercise is conducted in Russian, even for astronauts for whom it isn't their mother tongue. When NASA astronaut Karen Nyberg was training on the simulator for her flight to the International Space Station in 2013, she was asked what the biggest difference was compared to when she'd flown on a Space Shuttle in 2008. Her answer? "The Russian language—it's been a challenge for me," she said. "Hopefully, I've learned it to a point where I can be an efficient and effective crew member." When asked what language would be used on the mission, her commander Fyodor Yurchikhin replied "Renglish"—a unique blend of Russian and English that he said was exclusive to astronauts.

Despite the name, the descent module is also the place where astronauts sit during launch. Located inside a giant room that looks like a high-school gymnasium, you have to walk past a space toilet-training module to reach the seven steps that lead up to the simulator's hatch. Inside the cramped module, you're confronted with an intricate series of knobs, dials, lights, and screens. A bound manual breaks down every possible procedure for you.

Instructors often former cosmonauts—put trainees through their paces by introducing "off nominal situations"—emergencies, faults, and failures that need to be addressed. Of particular importance are those that could signal the need for an emergency landing. It could be that oxygen levels are rising, presenting a fire risk on re-entry. Under those circumstances, normal landing is aborted in favor of a quicker descent. Another serious problem might be a sudden depressurization of the capsule.

Dealing with fires is a major part of the training and it can be simulated, too, with smoke released inside the capsule by instructors. The first thing astronauts do is close the visors of their helmets to prevent smoke inhalation. Astronauts learn how to put out the fire in various ways, but there is no fire extinguisher or other fire-fighting equipment on board the Soyuz. We'll see how astronauts deal with fires in more detail on page 93. At the end of their training, the astronauts spend four days in the simulator running through all of these scenarios before taking an exam to check their readiness for the real thing.

↗ Chris Hadfield (left), Roman Romanenko (center), and Tom Marshburn (right) outside the Soyuz simulator at Star City.

> The Soyuz TMA-7 flying in space back in 2006. It has room for a crew of three.

THE SOYUZ SPACECRAFT

The workhorse of the Russian Space Agency (Roscosmos), the Soyuz spacecraft has been in constant operation for more than fifty years, making it the longest-serving human spacecraft in the world. The modern version is called the Soyuz TMA and is split into three sections: the instrumentation and service module at the base; the descent module above that; and, finally, the orbital module on the top. The whole thing weighs nearly 7.7 tons (7,000 kg).

The three launching astronauts sit inside the descent module, which is just under 7 feet (2.1 m) high and 7 feet 4 inches (2.2 m) across. With space at a premium, they sit side by side in reclining seats with their knees pressed up close to their chests. The flight engineer occupies the left-hand seat, and the commander is in the middle, behind the periscope and manual control sticks, with a second flight engineer (or sometimes a paying space tourist) on the right. To make room, the commander's seat is depressed lower than the other two, meaning that they often have to operate the control panel by using a stick to reach it. Alternatively, the flight engineer punches in the necessary sequence of buttons, before hitting "execute" on the say-so of the commander.

↘ ESA astronaut Thomas Pesquet (far) and Russian cosmonaut Oleg Novitsky (near) training inside the Soyuz simulator.

v An international team of astronauts gathered at the European Astronaut Centre (EAC) in Cologne, Germany.

Every astronaut candidate for the European Space Agency reports to the EAC in Cologne, Germany, for 16 months of basic training. Before they get into the finer details of space flight, they first learn about the history of space agencies around the world and are schooled in the relevant areas of space law (see p. 112).

Then the hard work really starts. As part of their basic training, ESA astronauts are expected to study diverse and complicated areas across various scientific disciplines, including electrical engineering, orbital mechanics, astronomy, and human physiology. They move on to learning specifically about the inner workings of the International Space Station (ISS) and the challenges of life on board, including guidance, navigation, and control. The final part of basic training includes instruction on rendezvous and docking, learning Russian, and acquiring basic scuba skills ahead of moving on to underwater training for spacewalks (see p. 20).

Once basic training is complete, advanced training takes another year. This training focuses more on hands-on work instead of classroom tasks and includes learning to fly the Russian Soyuz spacecraft and operating experiments in ESA's Columbus module. At this stage, astronauts mix with other nationalities from other space agencies' training centers.

YOU WAKE UP FILLED WITH NERVOUS EXCITEMENT. TODAY IS THE DAY YOU GET TO GO TO SPACE AND REALIZE A LIFETIME'S AMBITION.

You find yourself in the deserted steppes of Kazakhstan at the Baikonur Cosmodrome, the world's largest space launch facility and a place soaked in history. It was from here that Sputnik 1—Earth's first artificial satellite—and Vostok 1—the mission carrying Yuri Gagarin into orbit—set off. With the retirement of NASA's Space Shuttle program, it now sees all launches to the International Space Station (ISS). Russia leases the site from the Kazakh government, in a deal set to run until at least 2050 and worth $115 million a year.

Astronauts are a nostalgic bunch and partake in specific rituals associated with launching from Baikonur. Since the early 1970s, every astronaut has watched a Russian Western movie, *The White Sun of the Desert*, the night before lift-off. The next morning you sign your name on your bedroom door as you make your way to the launchpad. Your arrival is greeted with cheerleaders waving gold pompons. Don't expect, however, to see your rocket rolled out on the pad—this is considered bad luck. The rocket will be positioned ahead of your arrival, before being blessed by a Russian Orthodox priest. You'll walk an avenue of trees, with each one representing a previous

↗ Frank De Winne (ESA), Roman Romanenko (Roscomos), and Robert Thirsk (CSA; left to right) plant trees ahead of launch.

> A Russian Orthodox priest blessing the Baikonur launch site ahead of lift-off.

The Saturn V

Used for a brief flurry of activity between 1967 and 1973, the mighty Saturn V remains the tallest and heaviest rocket ever launched. At nearly 364 feet (111 m) tall, it would have towered over both the Statue of Liberty in New York and Big Ben in London. Launched a total of 13 times, it was mainly used to send astronauts to the Moon, but its last act was to send the US Skylab space station into orbit.

The Saturn V went from design to launch remarkably quickly. Plans started in January 1961, and the rocket was flying by November 1967. Flight technology had developed so quickly that the Saturn V burned through more fuel in one second than the total amount used by aviator Charles Lindbergh for the first solo crossing of the Atlantic in 1927.

The rocket may not have launched for nearly 50 years, but in 2013 NASA rocket scientists salvaged parts from museums to conduct a twenty-second hot-fire test at NASA's Marshall Space Flight Center. Their aim was to capture new data that they could feed into the design of modern rockets. Amazon billionaire Jeff Bezos has even funded missions to salvage parts of the Apollo 11 rocket from the ocean floor.

> The Saturn V rocket lifting off on July 16, 1969, carrying the Apollo 11 astronauts to the Moon.

mission to launch from Baikonur, and you'll plant your own sapling to keep up the sequence.

In an even more unusual tradition, astronauts tip their hats to Yuri Gagarin in a much more perverse way. Gagarin apparently asked the bus to stop en route to the launchpad so he could urinate. He duly relieved himself on the back right tire. Male astronauts repeat the exercise to this day, with some female astronauts bringing a prepared sample to splash on the tire instead. "Number twos" are less of a problem—this is because astronauts are offered enemas on the morning of launch.

You'll enter the Soyuz capsule two-and-half hours before the scheduled launch time. It is stowed away in the nose cone of the 165-foot (50-m), three-stage rocket, which contains about 330 tons (300 tonnes) of fuel. Four green mechanical arms hold it in place and will rotate away as the rocket launches. Music—often consisting of Russian love songs—is pumped into the cockpit during your final checks. Apparently, Gagarin asked for music on that maiden flight and the tradition persists.

The time has now come for you to leave Earth. Mission control counts you down in Russian: пять, четыре, три, два, один—and you're off. Pressed down in your seat, you soar into the sky toward the Kármán line and your first taste of outer space.

∧ The Soyuz rocket launching from Baikonur on July 14, 2012, carrying Expedition 32 to the ISS.

∨ ESA astronaut Samantha Cristoforetti signing the door of her bedroom on the morning of launch.

THE CHALLENGER DISASTER

Space launches have always been a high-risk endeavor. Nothing crystallizes the danger like a fatal accident, and the Challenger disaster in 1986 was a tragedy that halted our astral efforts for years.

On a sunny, but cold, January morning, the Space Shuttle climbed into the Floridian sky. But seventy-three seconds into the flight it broke up, witnessed in horror by onlookers, including the astronauts' friends and family. Millions more were watching live on television. All seven of the crew members were lost.

A subsequent investigation, driven forward by Nobel Prize-winning physicist Richard Feynman, eventually pointed the finger at a faulty O-ring (a fuel tank seal). It had cracked overnight in the frigid temperatures. Pressurized gas leaked during the launch, and the shuttle exploded when that compromised the external fuel tank.

The Space Shuttle fleet was grounded for nearly three years as the craft was redesigned to prevent the same thing from happening again. In 1988, the International Astronomical Union named seven craters on the far side of the Moon after the fallen astronauts (Jarvis, McAuliffe, McNair, Onizuka, Resnik, Scobee, and Smith). There is also a memorial in Arlington National Cemetery in Virginia, where some of their remains are buried.

v The crew of the Space Shuttle Challenger, ahead of their ill-fated mission in 1986.

MISSION CONTROL

ASTRONAUTS ARE THE ONES IN THE LIMELIGHT, BUT THEY'D BE LOST WITHOUT THE DEDICATED TEAM OF SPACE-FLIGHT EXPERTS ON THE GROUND KEEPING THEM SAFE.

These engineers and technicians monitor every aspect of a mission, from spacecraft systems to crew health.

The far wall of NASA's Mission Control Center in Houston, Texas, is covered in digital screens showing maps of the International Space Station's (ISS's) trajectory and live feeds from inside the station. Busy mission controllers sit at desks behind yet more screens, carefully poring over the latest data. On the walls are the plaques of previous missions controlled from the room, and those of the lost Challenger, Columbia, and Apollo 1 missions sit in the memorial area as a reminder of what's at stake.

< The patch of the lost Apollo 1 mission serves as a reminder to mission controllers.

> The Mission Control Center at the Johnson Space Center during the Space Shuttle era.

HOUSTON

The Texan city's name appears in some of the most famous words ever spoken in space. Apollo 11 astronaut Neil Armstrong said, "Houston, Tranquility Base here. The Eagle has landed." Apollo 13 Command Module Pilot Jack Swigert said, "Okay, Houston, we've had a problem" (not "Houston, we have a problem," as is commonly misquoted, thanks to Ron Howard's 1995 movie).

The Lyndon B. Johnson Space Center in Texas has been the home of NASA's Mission Control room since the Gemini 4 mission in 1965. Before that, it was based at Cape Canaveral in Florida. Today, flight controllers monitor missions to the ISS 24/7. A backup is located at the Marshall Space Flight Center in Alabama in case Houston is unduly affected by a hurricane or another emergency.

The original Apollo control room was decommissioned in 1992 and fell into a state of disrepair. In 2017, however, a Kickstarter campaign was launched to raise funds to restore it to its former glory. Replicas of the wallpaper and carpets will be installed. Even the original ashtrays and coffee cups will be brought back in time to celebrate the fiftieth anniversary of the first Moon landing in 2019.

NASA
Mission Control
esa
3D Attitude Display
FLIGHT ACTIVITIES
PAYLOAD DEPLOY/
RETRIEVAL (PDRS)

About fifty people each work nine-hour shifts, and the Center has been staffed twenty-four hours a day, seven days a week, every day of the year since 2000. There are similar rooms at the European Space Agency in Germany, the Russian space agency Roscosmos, and the Japanese space agency JAXA. Only about 10 percent of the mission controllers' time is spent actively controlling the astronauts and mission. It could be that they need to perform a burn to alter the trajectory of the ISS, or keep an eye on the station while the astronauts sleep. Most of the rest of the time (about 75 percent) is spent planning and organizing upcoming procedures. About 15 percent is spent on training.

∧ ESA astronaut Luca Parmitano is CAPCOM during a spacewalk in 2017.

Perhaps the most famous part of Mission Control is the "go/no go" sequence that used to be a familiar sound before launches during the Apollo and Space Shuttle eras. Just before the descent of Apollo 11 to the surface of the Moon, the following rang out around Mission Control. "Okay, all flight controllers go/no go for landing. Retro? Go. Fido? Go. Guidance? Go. Control? Go. Telecom? Go. GNC? Go. EECOM? Go. Surgeon? Go. CAPCOM we are go for landing." Flight director Gene Kranz (see box, opposite) was checking with each member of his team that everything looked good to proceed.

Here's a quick guide from NASA to some of the roles within mission control:

ROLES WITHIN MISSION CONTROL

ROLE	ABBREVIATION	RESPONSIBILITY
Flight Director	FD OR FLIGHT	Responsible for overall mission success
Spacecraft Communicator	CAPCOM	Communications link between flight control and astronauts
Flight Dynamics Officer	FDO	Pronounced "fido": plans maneuvers and monitors trajectory
Guidance Procedures Officer	GPO	Monitors onboard navigation and guidance computer software
Surgeon	SURGEON	A medical doctor on staff
Propulsion Engineer	PROP	Monitors reaction control and orbital maneuvering propellants
Guidance, Navigation, and Controls System Engineer	GNC	Monitors vehicle guidance and navigation systems
Electrical, Environmental, and Consumables Manager	EECOM	In charge of thermal controls of the vehicle, cabin atmosphere, supply systems, and fire detection
Instrumentation and Communications Officer	INCO	Monitors in-flight communications and instrumentation systems

Born in Toledo, Ohio, in 1933, Eugene Francis "Gene" Kranz would go on to become the most famous Flight Director in NASA's illustrious history. That's because the buck stopped with him during the Apollo missions to the Moon. He was in Mission Control when Neil Armstrong made those first historic steps off the foot of the ladder—as he was when the oxygen tank exploded aboard *Apollo 13* (see p. 59). He would later receive the Presidential Medal of Freedom.

Even at high school, the young Gene was obsessed with space travel, writing a thesis on "The Design and Possibilities of the Interplanetary Rocket." He went on to fight in the Korean War, returning home to work for the McDonnell Aircraft Corporation before moving on to NASA under the guidance of then-Flight Director Christopher Kraft.

He used his experience of losing colleagues in Korea after the Apollo 1 tragedy, in which a fire on the launchpad killed astronauts Gus Grissom, Ed White, and Roger Chaffee. "[I'd] never seen a . . . group of people . . . so shaken in their entire lives," Kranz said. He called his flight controllers together and told them: "You've got to get through this. You can't avoid it. The fact is that someone has died."

v Gene Kranz (second left) and his flight controllers celebrating the successful recovery of Apollo 13 in 1970.

TRAINING

ROCKET SEPARATION

WITHIN SECONDS YOU'RE CLEAR OF THE LAUNCH TOWER. JUST A FEW MORE MINUTES AND YOU'LL BE THERE—BUT NOT ALL OF YOUR ROCKET WILL MAKE THE TRIP WITH YOU.

Modern rockets are built in several stages, each designed for a specific part of the journey. By getting rid of parts of the rocket once they have been used, you lighten the mission and reduce the amount of fuel needed. This can be explained by Isaac Newton's second law of motion. The less mass you have, the less force you have to apply to achieve the same acceleration.

Rockets, such as the Soyuz used today, are the distant descendants of the first multistage rockets launched by space-flight pioneers Robert H. Goddard in the 1920s (see box, opposite) and Wernher von Braun in the 1960s (see box, p. 46).

At the base of the Soyuz rocket is a core engine surrounded by four boosters, and together these five engines do the initial heavy lifting. In less than a minute, you are higher than Mount Everest and experience G-forces close to 2 g as the rocket executes a pitch maneuver to achieve the correct approach angle for the desired orbit. After two minutes, you're above 25 miles (40 km), high enough for the sky to begin turning black. Here, the Launch Escape Tower is discarded—the emergency evacuation module that would have saved you from a

> The Soyuz capsule in orbit around Earth.

↲ The solid rocket boosters from a Space Shuttle being recovered from the Atlantic Ocean and towed back to Florida.

ROBERT H. GODDARD (1882–1945)

v Robert Goddard greatly improved the design of early rockets and is remembered as the father of American rocketry.

On March 16, 1926, the world's first liquid rocket rose into the air above Auburn, Massachusetts. Its designer, Robert Goddard, is widely lauded as the father of American rocket science. The flight lasted just two-and-a-half seconds, reaching an altitude of 41 feet (12.5 m) before careering into a cabbage patch 184 feet (56 m) away. Within half a century, liquid, multistage rockets were carrying astronauts all the way to the Moon.

Goddard was awarded a total of 214 patents, including those relating to liquid and multistage rockets. His obsession with breaking free of gravity's shackles started early. At 16, he devoured H. G. Wells's science fiction story *War of The Worlds*. A year later, he climbed a cherry tree, daydreaming of building a device that would allow him to keep on ascending all the way to Mars.

His early work on rockets gained financial support from The Smithsonian Institution, who awarded him a $5,000 grant in 1917. By 1930, Goddard's rockets had shown such promise that financier Daniel Guggenheim gave him $100,000 (almost $2 million in today's money). In recognition of his landmark contributions to the field of rocketry, NASA's Goddard Space Flight Center in Maryland is named after him, as is a crater on the Moon.

Von Braun's name is synonymous with the mighty, multistage Saturn V rocket that launched the Apollo astronauts, but in the 1930s and 1940s he worked for Hitler and the Nazi Party in Germany. He was instrumental in building the V2 rockets that rained down on London during World War II. His V2 was the first ever object to cross the Kármán line and enter space, on October 3, 1942.

When it was apparent that Germany would lose, he engineered the situation so that he could surrender to the Americans instead of the Soviets, and he was relocated State-side under an American military intelligence program known as Operation Paperclip. The V2 was used by the United States to put the first living things into space in 1947 (see pp. 64–67), and von Braun worked on better rockets.

The engineer was enamored with space and rockets from a young age. He owned a telescope, and at 12 years old, he was taken into police custody when he let off fireworks attached to a toy wagon. Clearly his future career was set. He maintained that he'd always wanted rocket technology to be applied to peaceful space exploration, reportedly saying to a colleague after a V2 hit London, "The rocket worked perfectly, except for landing on the wrong planet."

v German Wernher von Braun came to the United States after the fall of the Nazis in 1945.

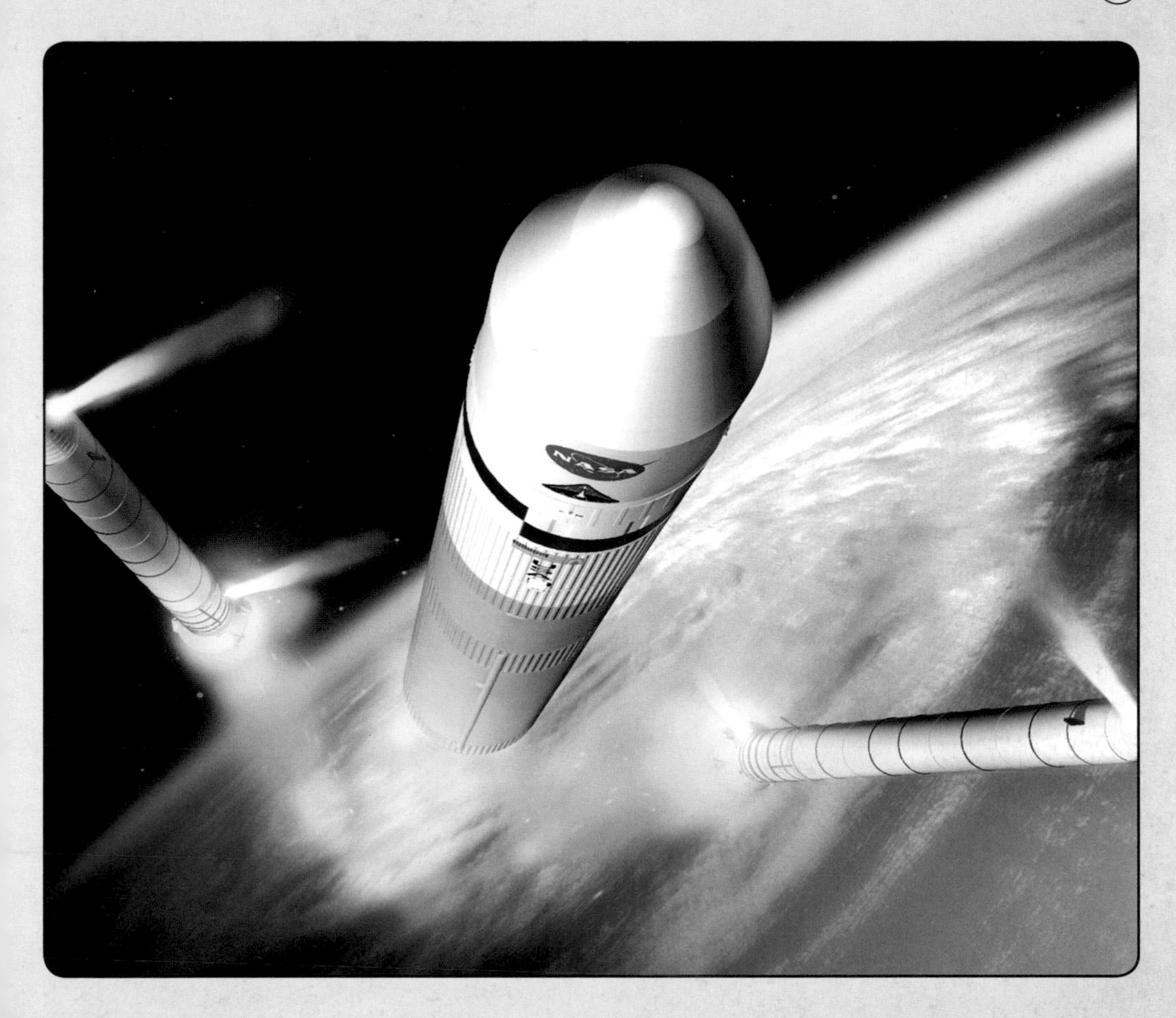

faulty launch. If anything goes wrong now, however, you can still separate from the rocket and make a safe reentry and landing.

Next, the four spent rocket boosters fall away and head back to Earth, landing in the middle of the desert some 217 miles (350 km) away from the launchpad. You've now entered the second stage, during which the core engine continues to fire for around three minutes. Meanwhile, the nose cone is jettisoned, exposing your Soyuz capsule for the first time and giving you a view out of the window. The panorama is jaw-dropping, although the G-forces restrict your ability to take a peek.

The second stage ends when you've already technically become an astronaut, at an altitude of 105 miles (170 km). The core engine switches off as the back of the rocket separates and again drops back down to the dry desert below. The third-stage engine fires up, and over the next four minutes you're carried up to your intended orbit at 137 miles (220 km) above Earth. The whole procedure takes only nine minutes from lift-off. As the final engine switches off, the capsule falls eerily silent. The Soyuz's solar panels and communication arrays deploy. You notice the cuddly toy mascot hanging from the instrument panel in front of you floating freely, the first sign that you've entered weightlessness and are in freefall around Earth.

∧ An artist's impression of the solid rocket boosters separating from the canceled NASA Ares V rocket.

WOULD YOU BE A "LEAD HEAD"? THAT'S THE MONIKER EARNED BY ABOUT ONE-QUARTER OF ASTRONAUTS WHO DON'T SUCCUMB EASILY TO SPACE SICKNESS. THE REST, UNFORTUNATELY, ARE NOT SO LUCKY.

Spending all your time on Earth, it is easy to forget how well evolution has adapted your body to the rigors of life on a planet. Head into space, however, and your body quickly lets you know you've pushed it far from its comfort zone. Without your usual gravitational field to tell you which way is up, weightlessness soon sends the vestibular system in your inner ear haywire. This system, responsible for balance and your sense of spatial orientation, comprises a liquid that shifts around as you move and stimulates tiny hairs to send signals to your brain. In a weightless environment, the liquid floats around in the vestibular tubes, triggering the hairs when you're not actually moving. Conflicting signals from the hairs and your eyes lead to most astronauts experiencing nausea, visual illusions, and disorientation soon after entering weightlessness. Some even vomit (see box, right, and see box, p. 51), and that carries with it the risk of dehydration.

The scientific name for this space sickness is space adaption syndrome (SAS). It doesn't always follow that those who cope well training on the Vomit Comet (see pp. 28–31) escape the effects of SAS when in space. The good news is that the symptoms normally pass after two to four days as your body adjusts to a new normal. The bad news is that you often have to ride out the nausea. Antisickness medications are usually avoided because they make space travelers drowsy. The exception to this is during extra-vehicular activities (see pp. 80–83).

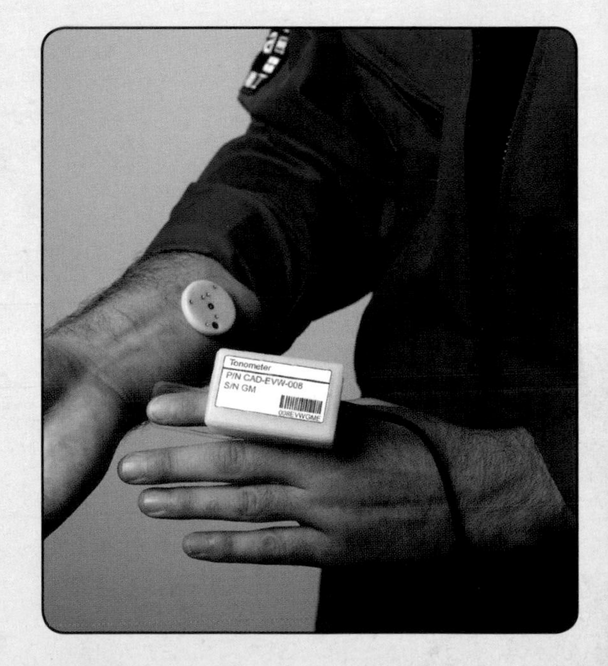

∧ EveryWear is a prototype device developed by France's space agency CNES to monitor astronauts' health.

↗ Sled, an experiment that investigated space sickness, involved monitoring eye movement while cold or hot air was blown into the astronaut's ears.

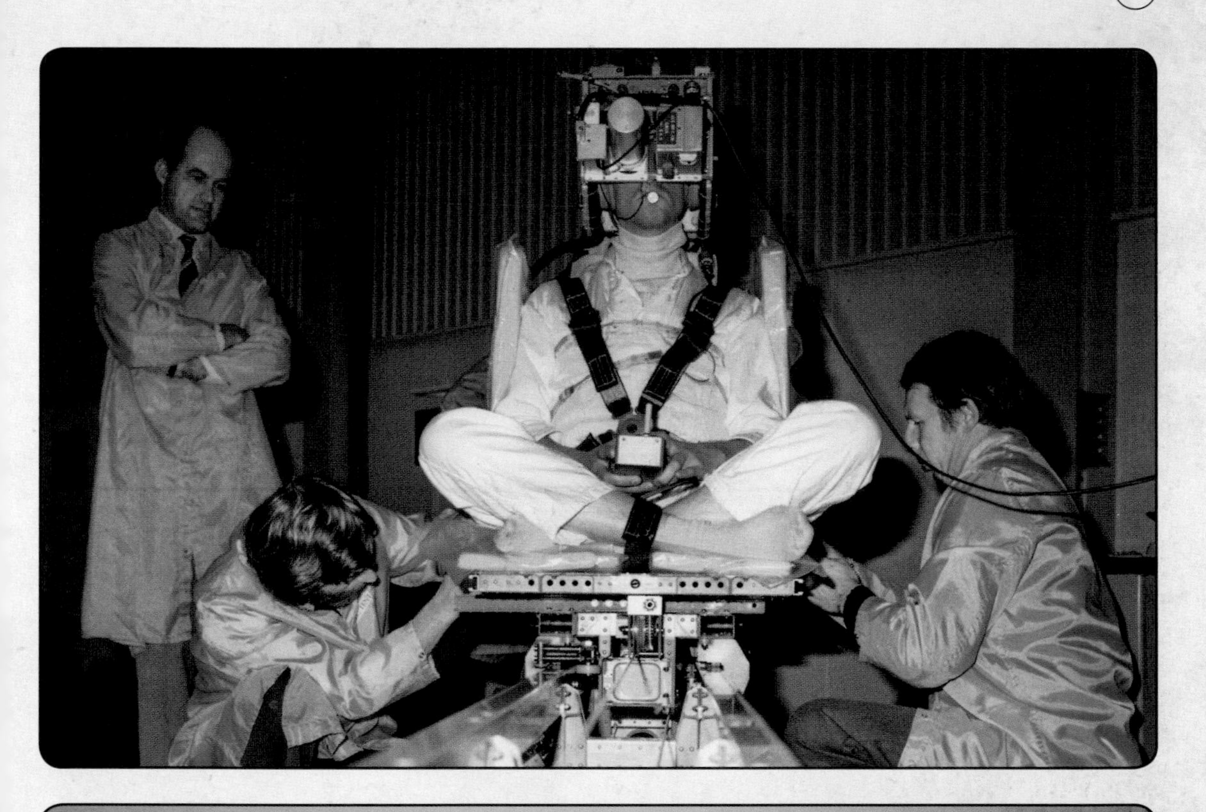

THE GARN SCALE

Poor Jake Garn. When the former senator for Utah launched into space on the fourth flight of the Space Shuttle Discovery in 1985, he might have secretly hoped he'd make his own little piece of history. He certainly did, however, probably not in the way he'd originally pictured. He had secured his place on the shuttle through his position on the Senate Appropriations Committee and flew as a government observer and payload specialist. He might be the first sitting member of Congress to fly into space, but he's now more remembered for his delicate constitution.

Garn apparently fell foul of space sickness more than any astronaut before or since. The rumor is that his crew mates had to Velcro him to a wall for the duration of the mission, which lasted 167 hours and completed 108 orbits of the planet. Garn himself is tight-lipped, but he admits he did feel nauseated.

In recognition of his unusual accolade, NASA mission controllers unofficially use the Garn Scale to rank how badly an astronaut is affected by space sickness. Garn himself sits at one Garn. Most astronauts rarely register one-tenth of a Garn.

v Senator-turned-astronaut Jake Garn during his uncomfortable visit to space.

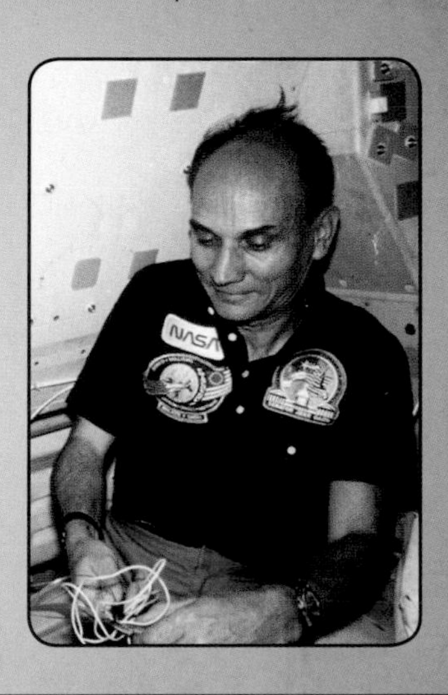

Astronauts on these space walks wear medicated patches to combat sickness, because vomiting in your spacesuit can be lethal. The vomit droplets would float around inside your helmet, putting you at risk of drowning in it.

Space stations are also designed to help you adjust. The modules of the ISS all share matching orientations, and the writing on the walls always appears the same way around. This helps create an artificial sense of up and down. Newbie crew members are told to orientate themselves in the same direction as the words and their fellow crew members. They are told to avoid sudden head movements and to resist attempts to indulge in microgravity gymnastics until their vestibular system settles down. If space tourism takes off in the coming years (see pp. 124–135), there will be a burgeoning business opportunity to provide ways of combating SAS in tomorrow's astronauts, whose time in space will be shorter than the typical adjustment periods that today's space travelers experience.

> ESA astronaut André Kuipers preparing samples for the Integrated Immune Study, which is looking into changes to astronauts' immune systems during space flight.

v Astronaut Clayton Anderson with a floating water bubble on the Space Shuttle Discovery. Vomit acts in a similar way.

Imagine someone throws up in a weightless environment. Tiny droplets of vomit would be constantly floating around. That's not just disgusting—it's also a potential danger to the spacecraft's intricate electrical systems. Back in 2013, during a Q&A with a school, International Space Station Commander Chris Hadfield explained the solution to the problem of astronauts being sick. They use barf bags to make sure they catch all of it.

The plastic bags are lined with paper so that you can wipe your mouth and face after vomiting. You push the vomit down into the end and seal it tightly with a ziplock at the top to prevent it from leaking. "What am I going to do with it?," Hadfield asks. "This bag has to stay with me in space for many months." Once sealed, the sturdy bag is deposited in the wet trash.

Barf bags and other wet trash containers are tested at the Advanced Water Recovery Systems Development Facility at the Johnson Space Center in Houston, Texas. Scientists there create fake vomit to check that the bags adequately contain it. In case you were wondering, they make it from pureed cottage cheese, tomato soup, apple juice, soy sauce, and frozen vegetables.

v Mock-up of the ECLSS at the Marshall Space Flight Center in Alabama.

TRAINING

Rendezvous and Docking

THE SUN COMES UP, FLOODING YOUR SOYUZ CAPSULE WITH LIGHT. OUT OF THE WINDOW YOU CAN SEE THE BRILLIANT BLUE OF THE EARTH BELOW AS THE PLANET CURVES AWAY BENEATH.

The engines have fallen silent, and your seat belt is the only thing stopping you from floating around and enjoying weightlessness.

However, getting yourself into orbit around Earth is only half the battle. Now you need to catch up with the International Space Station (ISS). Originally, this rendezvous maneuver took two days. Fortunately, since 2013, it has been shortened to just five or six hours. Soon, you'll be greeting the astronauts already aboard the ISS, but currently you're at an altitude of only 137 miles (220 km). The ISS orbits at around 261 miles (420 km). Jumping between the two orbits cannot be done in a single maneuver; it takes two stages.

The first step is called a Hohmann Transfer, after the German scientist Walter Hohmann (1880–1945), who first devised it. The engines are fired once to climb to a middling altitude called the phasing orbit. A second engine burn keeps the Soyuz at the right speed to stay there. It is called the phasing orbit, because it is used to decrease the angle—or phase—between you and the ISS. Initially, you'll be busy with capsule checks, but when these are complete, you'll have your first chance to relax. You can unclip your seat belt, loosen your helmet, and remove your gloves. You can even float up into the Orbital module above to make use of the space toilet. At this point, you'll experience five so-called "deaf orbits," when you're out of radio contact with ground control.

∧ The Russian Soyuz 39 spacecraft (foreground) and *Progress 55* spacecraft docked to the International Space Station.

↗ The Space Shuttle Endeavour docked to the ISS, seen from the Russian Soyuz spacecraft.

GEMINI 8

The first docking in space nearly ended in disaster. On March 16, 1966, NASA launched the *Gemini 8* mission, carrying astronauts Neil Armstrong and Dave Scott into orbit. Almost two hours earlier, they'd also sent up the uncrewed Agina module.

The plan was to have *Gemini* dock with the Agina—a space first and a crucial step in preparations for a moon shot. Things went smoothly to begin with, but quickly turned bad. A program in the Agina's computer memory sent the conjoined spacecraft spinning the wrong way. To make matters worse, Armstrong and Scott were temporarily out of range of ground communications.

Armstrong undocked, but that caused *Gemini 8* to start spinning wildly, completing one revolution every second. The astronauts' vision started to blur. Much longer and they'd have passed out. Armstrong's quick thinking saved their lives. He turned everything off and used the reentry thrusters to stabilize the craft.

With the unthinkable averted, mission control ordered them to return to Earth immediately. Landing more than 600 miles (965 km) south of Japan, they were rescued by the US Navy destroyer USS Leonard Mason. Both Armstrong and Scott would later walk on the Moon as part of the Apollo program.

^ The uncrewed Agina seen by Armstrong and Scott through the window of Gemini 8.

Getting from the phasing orbit to the altitude of the ISS requires three engine burns during what space scientists call a bielliptic transfer. The first two burns get you up there; the third keeps you traveling at the right speed to approach the ISS. Rendezvous with the space station is then a completely automated process, but the crew constantly checks to make sure it's progressing smoothly. The commander is always poised to take over manual control if something goes wrong.

As the Soyuz approaches the ISS, a probe is deployed and guided into a cone on the space station, which closes around it. The two spacecraft are drawn together, and eight hooks click into place to create a tight seal. However, you can't enter the ISS just yet—the crew will need to perform one to two hours' worth of checks to make sure the seal is sufficiently tight. Only then can you knock on the hatch and wait to hear the astronauts inside knock back. The hatch is then opened, and there are hugs all around as you get to know your new home.

∧ The Soyuz, with its outstretched probe, about to dock with the ISS back in 2011.

Until recently, all dockings in space were carefully choreographed from the ground or by astronauts on board. However, 2007 saw the first completely autonomous joining of two spacecraft in orbit.

At 311 miles (500 km) up, the Autonomous Space Transport Robotic Operations (ASTRO) craft separated from the NextSat satellite until they were 33 feet (10 m) apart. After a ninety-minute orbit, their computer systems reacquainted them without any input from Mission Control. This "autopilot" technology will be crucial for refueling and repairing aging satellites in orbit.

In February 2017, NASA took another step toward making autonomous docking the norm when it launched the Raven experiment to the ISS on board a SpaceX rocket. Perched on the ISS's exterior, it monitors all incoming and outgoing spacecraft and collects valuable data on their speeds and distances. The information is fed into algorithms designed to replicate the docking without manual input.

As we expand our activities in space, pushing our machines ever farther from Earth, autonomous docking becomes the only option for multistage missions. The time it takes for a signal to travel between Earth and Mars, for example, is typically twenty minutes, so real-time operation from Mission Control back home becomes impossible.

∨ Orbital Express NextSat photographed in space by the Orbital Express ASTRO satellite.

PART 2

LIVING IN SPACE

- BREATHING
- EATING AND DRINKING IN SPACE
- ANIMALS AND PLANTS IN SPACE
- EFFECTS OF MICROGRAVITY
- EXERCISE
- HOW TO SHIELD YOURSELF FROM RADIATION
- SPACEWALKS
- SLEEP
- WASHING AND CLEANING
- EMERGENCIES
- TOILETS
- GENERATING POWER
- PSYCHOLOGY
- SPACE JUNK
- SPACE LAW
- REENTRY

BREATHING

RUNNING OUT OF OXYGEN IS ONE OF THE THINGS THAT WILL KILL YOU THE QUICKEST, SO SCIENTISTS AND ENGINEERS HAVE INVENTED SEVERAL WAYS TO KEEP SPACE TRAVELERS BREATHING.

It's all down to chemistry and thriftiness—you want to make oxygen as easily as possible and reuse everything you can to keep costs down.

Launching oxygen tanks from Earth is expensive, so it's restricted to only backup supplies. Astronauts aboard the International Space Station (ISS) largely breathe the same mix of air as we do here on Earth, but their oxygen is generated through a chemical reaction called electrolysis. Electricity from the ISS's huge solar panels (see pp. 120–121) is passed through waste water on board. That pries apart the two hydrogen atoms and one oxygen atom that make up water's molecular H_2O structure. The hydrogen was originally vented overboard, but now there's a clever way to use it by adding it to the carbon dioxide that astronauts breathe out (see box, opposite). Hydrogen (H_2) and carbon dioxide (CO_2) are combined in the ISS's Sabatier System to create water (H_2O) and methane (CH_4). The water can then be used to make more oxygen. The methane is vented into space.

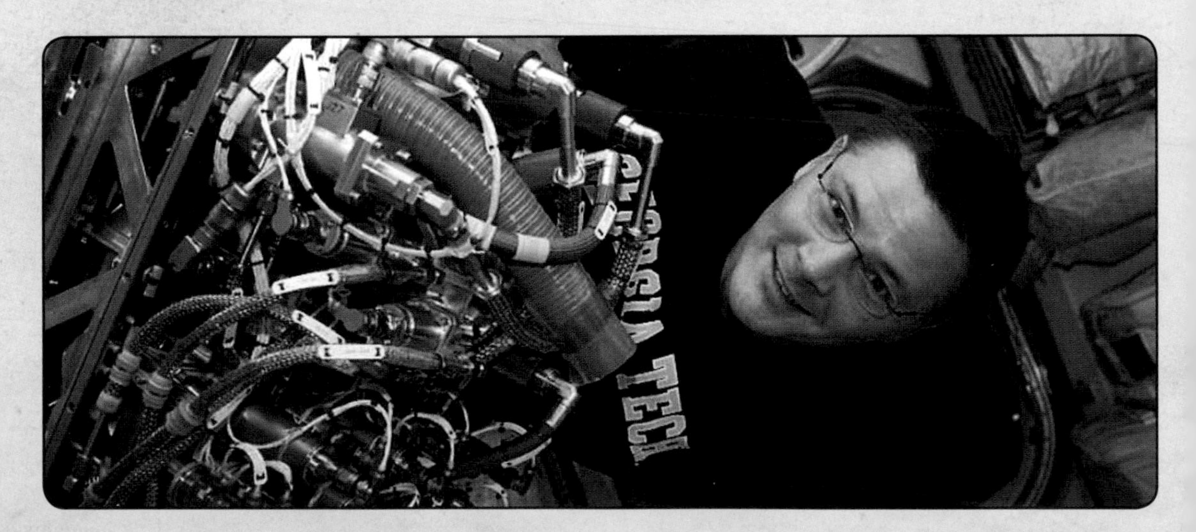

SCRUBBING CARBON DIOXIDE

On Earth, your cells combine oxygen from the air with glucose from food to produce energy, carbon dioxide (CO_2), and water for your body. You then breathe CO_2 out, because it's dangerous for humans in high concentrations. Our atmosphere is just 0.04 percent CO_2, largely thanks to trees, plants, and microbes that take in the gas for their own use and substitute it for oxygen. If CO_2 levels rose above 1 percent, you would start to feel drowsy. At 8 percent you would start to shake, sweat, and potentially pass out. CO_2 concentrations above 8 percent are lethal.

In space, there are no natural CO_2 absorbers, so we take our own in the form of canisters of lithium hydroxide (LiOH). The CO_2 reacts with LiOH to produce lithium carbonate (Li_2CO_3) and water (H_2O). Removing carbon dioxide in this way is known as "scrubbing." When the canisters run out of lithium hydroxide, they need to be replaced. The US Destiny lab on the ISS uses an alternative method of CO_2 removal called the carbon dioxide removal assembly (CDRA). It relies on crystals called zeolites to sift out CO_2, which is then safely vented into space. Unlike the LiOH canisters, they don't need replacing.

v Astronaut Megan McArthur working with lithium hydroxide (LiOH) canisters from beneath Space Shuttle Atlantis's mid-deck.

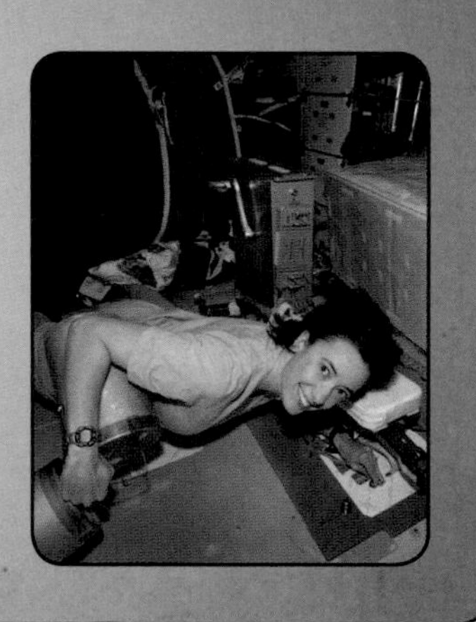

↲ NASA astronaut Doug Wheelock, Expedition 25 commander, works to install the Sabatier System.

v Canadian astronaut Chris Hadfield checks on an experiment to capture the oxygen from carbon dioxide.

There were previously two systems on board for making oxygen in this way: the Russian Elektron System in the Zvezda module and the US Oxygen Generation System in the Destiny module. However, the Russian system failed in 2005. The US system now goes through 6 gallons (23 liters) of water a day and can provide 12 pounds (5.4 kg) of oxygen every day for a crew of four or 20 pounds (9 kg) for six astronauts. The potential for failures means that there are backup methods to make sure there is always oxygen. That's exactly what the crew used when Elektron broke. Astronauts initially repressurized the cabin by releasing oxygen from tanks delivered by an uncrewed service module. They then tested the next line of defense: potassium perchlorate candles. When lit, these can produce six-and-a-half hours of oxygen for every kilogram, but they can also be dangerous. In 1997, a candle aboard the Russian Mir space station malfunctioned and caught fire.

Astronauts who venture outside the protective bubble of a space station must wear a pressurized space suit at all times (see pp. 80–83). What if you aren't wearing one? You might think holding your breath is best, but it would be catastrophic. No longer under any pressure, the air inside you would rapidly expand and rupture your lungs. You have only about fifteen seconds of oxygen in your body and without it you'll pass out. Bubbles will form in your blood—a condition called ebullism—and you'll be in excruciating pain as you swell up to twice your normal size. You'll be dead soon afterward.

∧ Cosmonaut Sergei Krikalev repairing the Elektron oxygen generator in the ISS's Zvezda service module.

It is one of the luckiest escapes in the history of space travel. Two days into a flight to the Moon in 1970, an oxygen tank exploded, damaging the service module of Apollo 13 and making a lunar landing impossible. All attention immediately turned to getting the crew home safely. With the service module stricken, all three astronauts moved into the lunar module—a home designed for two astronauts to spend a day and half. The trio would have to ride it out in there for four days if they were to make it back to Earth.

A major problem was the lack of sufficient lithium hydroxide (LiOH) to scrub the extra carbon dioxide exhaled by the additional astronaut. Frustratingly, the square-shaped LiOH canisters in the service module—of which there were plenty—were not compatible with the round canisters in the lunar module. As famously depicted in the 1995 movie Apollo 13, flight engineers fashioned a device they called "the mailbox" to deliver the square-shaped canisters to the round sockets, using a return hose from one astronaut's space suit. With the crew now able to continue breathing, they successfully used the Moon's gravity to catapult home safely.

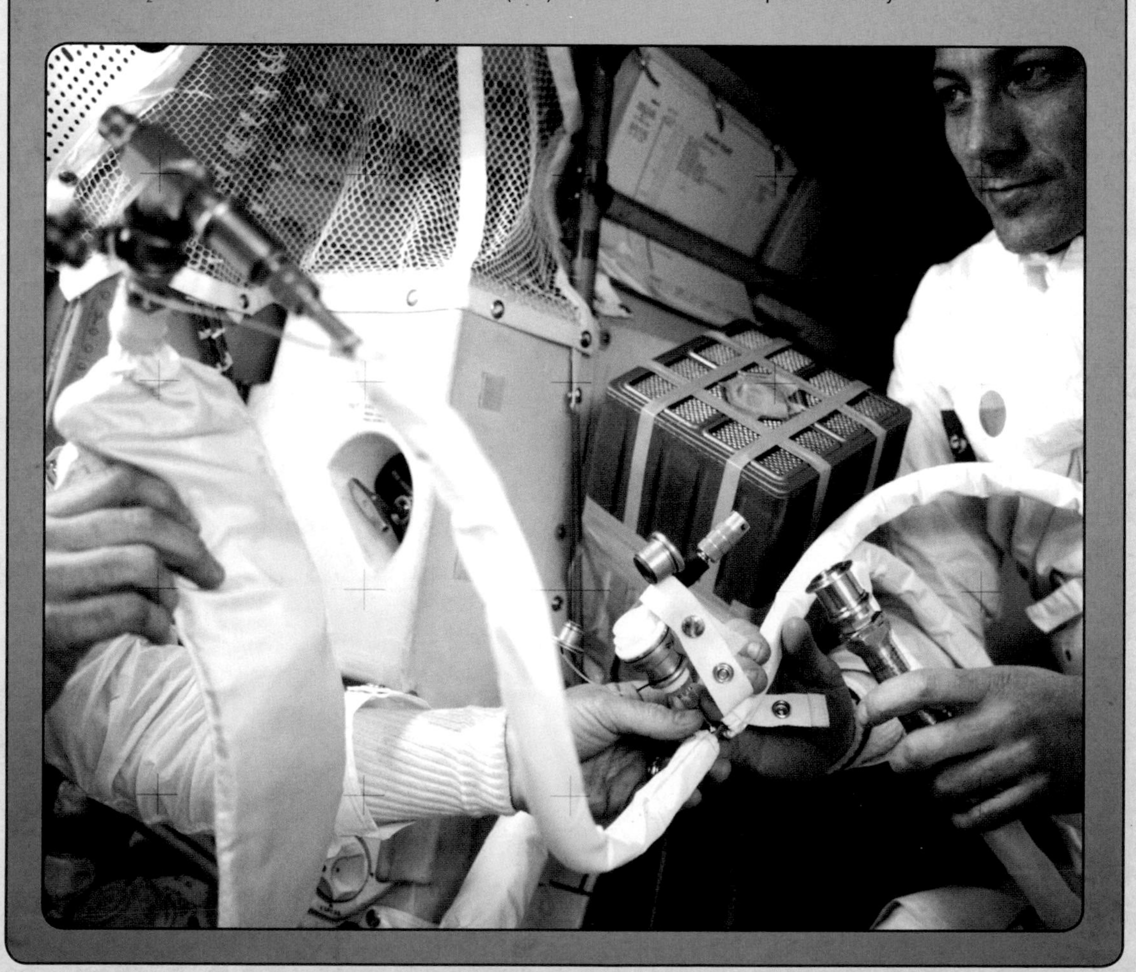

v The "mailbox" used by the Apollo 13 astronauts to scrub the extra CO_2 in the lunar module.

EATING AND DRINKING IN SPACE

WHILE ORBITING AROUND EARTH AT 17,500 MILES PER HOUR, JOHN GLENN'S ATTENTION TURNED TO HIS STOMACH.

The first American to orbit the planet was also the first US astronaut to eat in space. Before his flight, it was unclear whether humans could swallow in weightlessness, or even if their bodies could absorb the necessary nutrients.

Glenn's maiden meal was far from appetizing—pureed applesauce in an aluminum tube, which he had to suck through a straw. These early astronaut snacks were based on army rations. The food got slightly more adventurous as we expanded our space travel horizons, and inspiration was taken from the long history of exploration on Earth. Adventurers and explorers realized long ago that drying food was a great way to make it last, so a lot of space food is dehydrated or freeze-dried. This also cuts down on the weight at launch. Mission controllers recognize that home comforts can be a real boost to morale, so during the Apollo 8 mission to the Moon, the astronauts opened a meal of thermostabilized turkey, gravy, and cranberry sauce. It was Christmas Eve, after all.

These days, astronauts on the International Space Station (ISS) have life a little better. There are more than 100 items on the menu, and astronauts often select their meals on the ground ahead of launch. They eat three meals a day, along with snacks in between. The number of calories they consume is carefully monitored, as they fill out computerized questionnaires on their diet. Dieticians on the ground can give them advice if their intake needs tweaking.

Condiments, such as mustard, ketchup, and mayonnaise, are available, as are salt and pepper. However, those last two have to be in liquid form; powders are a real no-no, because they can float away and damage the ISS's intricate electronics or get into the astronauts' eyes. Anything that can create crumbs is also out of bounds for the same reason. That includes bread and cookies. Astronauts eat tortilla wraps instead, often with peanut butter.

You won't find carbonated beverages on the menu; microgravity causes them to lose their fizz. However, coffee, tea, and orange juice are available. Alcohol is strictly

∧ Peggy Whitson (left) and Valery Korzun (right) eating hamburgers on the International Space Station.

↗ Food prepared for astronauts on board the ISS, including a cheese spread and creamed spinach.

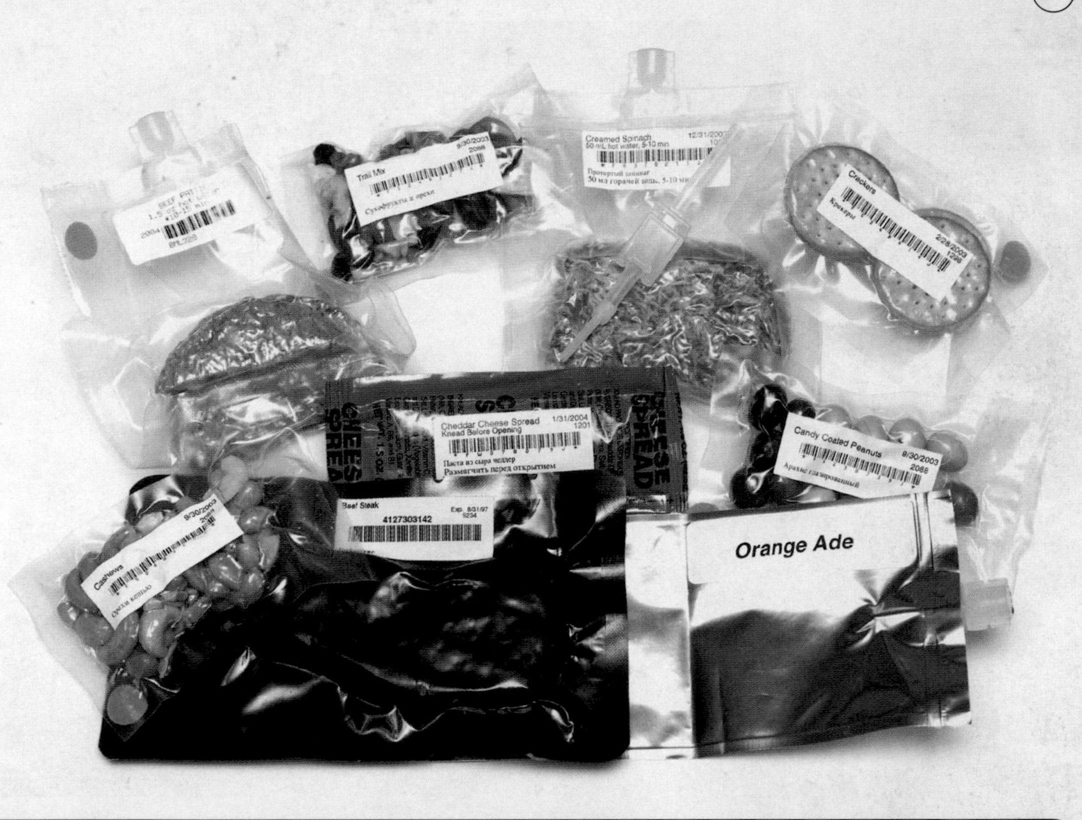

WHERE DOES WATER COME FROM ON THE ISS?

When it comes to drinking water, there is a fundamental difference between Russians and Americans on the ISS. US astronauts drink their own recycled urine, but their Russian colleagues don't.

In the US module, 93 percent of all water on board is recycled. This means processing urine, shower runoff, sweat, and even urine from any animals on board. It is all cleaned and purified, leading to a crisp, fresh taste that is apparently indistinguishable from bottled water.

While the Russians do recycle condensation from the air, they draw the line at urine. Instead, they collect it in bags and it's transferred to the US end for recycling. So the Americans end up drinking Russian pee. This saves an extra 1,585 gallons (6,000 liters) of water every year, cutting down the need to launch new water into orbit.

The water is cleansed using the Water Recovery System, which forms part of the Environmental Control and Life Support System (ECLSS). The same technology has been adapted for use on Earth and is now used in rural Mexico to provide clean drinking water to impoverished areas where parasites and stomach bugs are rife. It's just one of the examples of how space exploration benefits us all.

v NASA astronaut Leland Melvin with part of the Water Recovery System aboard the ISS.

v On the ISS, ESA astronaut Samantha Cristoforetti sips espresso from a cup designed for use in |zero gravity.

forbidden on the ISS, not only for its intoxicating properties but also because its presence in astronaut urine can affect how efficiently that water is recycled (see box, p. 61). Having said that, there have been boozy space trips before. There are plenty of stories of Russian cosmonauts smuggling tipples into space. Those visiting the now defunct Russian Mir space station were even permitted official rations of cognac and vodka. During the Apollo 11 mission, astronaut Buzz Aldrin sipped wine before stepping out onto the surface of the Moon, as part of a religious Communion.

v Peanut butter spread on a tortilla floating on the mid-deck of the Space Shuttle Discovery.

> British astronaut Tim Peake in his makeshift tuxedo as he samples Heston Blumenthal's food in space.

On the whole, space food is dull—whatever is quick, safe, and lets astronauts refuel efficiently. Yet, missing home comforts can take a toll on morale. So it was a boost to British astronaut Tim Peake's stay on the ISS in 2016 when some of his meals were cooked by Michelin-starred chef Heston Blumenthal.

Heston—noted for using chemistry in cooking—and Tim worked together for two years before launch, and based on their conversations, the chef devised seven meals that would remind Tim of life back on Earth with his family. Tim's taste buds were tested at Blumenthal's restaurant in Bray, England, to check his sensitivity to flavor, texture, acidity, and spice.

The seven dishes included beef and black truffle stew, a bacon sandwich (a British stalwart), and Alaskan salmon. That last one was based on a memory Tim shared about a fish he had caught and cooked on an open fire as part of his military training. The meals were canned and flew with him as part of his bonus food allowance. Some of the food was eaten live on British television, with Tim dressed up in a makeshift tuxedo. A far cry from John Glenn's applesauce!

ANIMALS AND PLANTS IN SPACE

IT'S FEBRUARY 20, 1947, AND A V2 ROCKET CONFISCATED FROM THE NAZIS STANDS ON THE LAUNCHPAD IN WHITE SANDS, NEW MEXICO.

Hidden away at the top of the rocket is a group of fruit flies, the first living emissaries to leave Earth behind. It would take them just three minutes and ten seconds to reach an altitude of 68 miles (109 km), more than enough to cross the Kármán line. Their trip jump-started a remarkable flurry of animal space activity designed as a way for humans to dip their toes into the unknown cosmic ocean. Could living things survive beyond the upper atmosphere? Would radiation from space kill them? The fruit flies were parachuted back to the ground alive and well.

During the next decade, the United States and Soviet Union became more ambitious with their animal astronauts. NASA launched a rhesus monkey named Albert II in 1949. His predecessor Albert I had failed to reach space. Albert II

SPIDERS IN SPACE

In 1973, NASA launched two eight-legged astronauts into orbit aboard the Skylab 3 space station. Sending the two spiders—named Arabella and Anita—into space was the brainchild of US high-school student Judy Miles, who wanted to know if a spider could spin a web in space.

The spiders initially struggled to adapt to the weightless environment, but within a couple of days were spinning their webs as normal. The strands were noticeably thinner, but the pattern was the same as on Earth. Their bodies are now preserved in the Smithsonian Air and Space Museum in Washington, D.C.

Spiders flew again on the Space Shuttle Endeavour in 2008. Students in Colorado and Texas had the same kind of spiders in their classroom, and used video and images from space to compare and contrast their webs. The space webs were more tangled.

A similar experiment was repeated in 2011 when two golden orb spiders—named Gladys and Esmerelda by astronaut Casey Colman—journeyed into space aboard the penultimate Space Shuttle mission. Both arachnids spun webs that were more circular than the terrestrial version. During their time in orbit, they snacked on a diet of fruit flies sent up as spider food.

Arabella, a common cross spider, spins the first web in space on board NASA's Skylab space station.

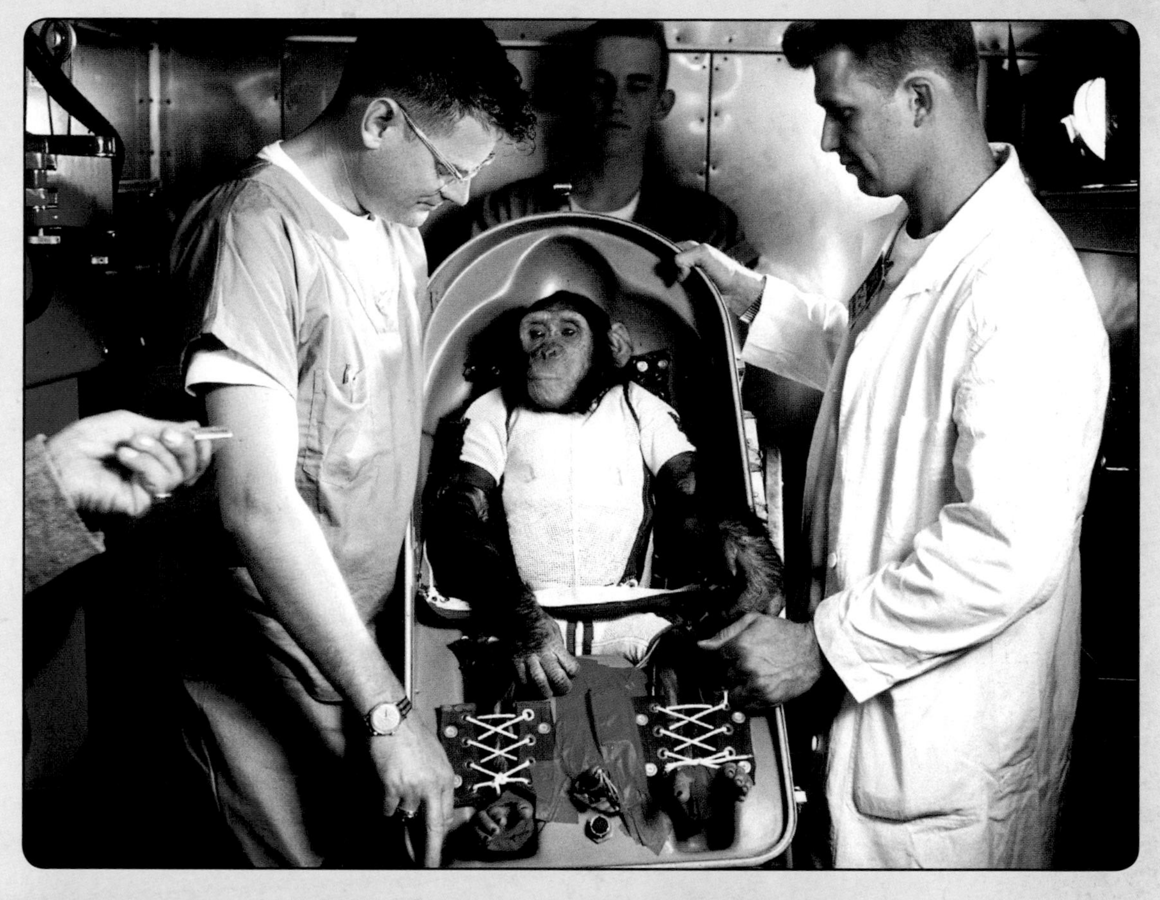

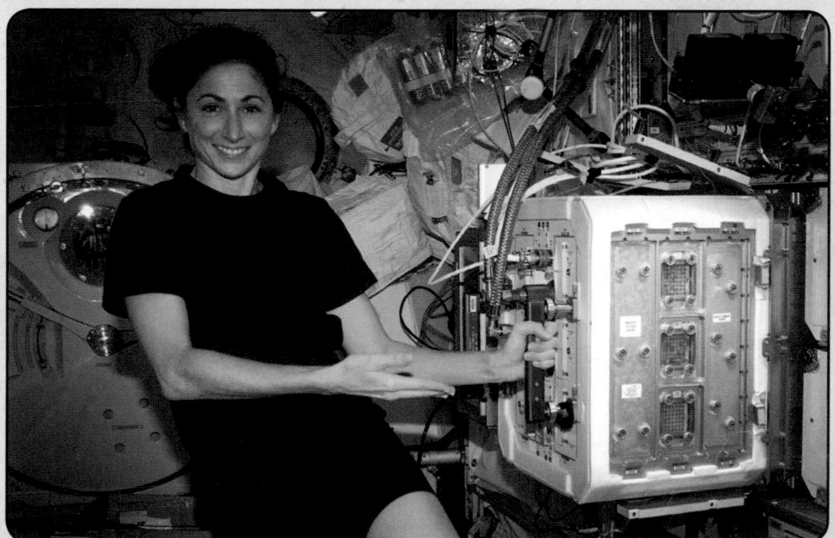

∧ Three-year-old chimpanzee Ham inside the protective pod, ready to be launched on a sub-orbital arc over 150 miles (241 km) high.

< NASA astronaut Nicole Stott pictured near the Mice Drawer System (MDS) in the Kibo laboratory of the International Space Station.

made it, but the parachute failed to deploy before landing and he died on impact. The Soviets turned to dogs in the 1950s, starting with Tsygan and Dezik. They got to space and survived the journey home, but never reached orbit. They were followed by Laika, who became the first living thing to orbit Earth in 1957. However, without the provision of a reentry mechanism, she quickly died. Her craft—Sputnik 2—burned up in the atmosphere six months later.

As part of their Mercury program, NASA used a chimpanzee named Ham, after the Holloman Aerospace Medical Center in New Mexico. When he was launched on a suborbital flight from Cape Canaveral in January 1961, it marked the end of a long road from his home in Cameroon, where he had been caught by trappers. He wore a space suit and had been trained to push a series of levers while in flight. After a trip lasting nearly seventeen minutes, his capsule splashed down into the Atlantic Ocean and he was recovered safely. His only injury was a bruised nose.

Without these early space pioneers, Yuri Gagarin and John Glenn would not have been able to make their landmark orbital trips in the early 1960s. Two tortoises would later fly around the Moon to test whether it was safe for humans to follow (see p. 136). In the decades since, an array of creatures have made it over the Kármán line to further enhance our understanding of the effects of space travel and microgravity on living organisms. They include mice, frogs, guinea pigs, rabbits, rats, cats, wasps, beetles, fungi, spiders (see box, p. 64), fish, newts, crickets, snails, sea urchins, shrimp, jellyfish, butterflies, scorpions, cockroaches, bees, ants, and geckos.

v NASA astronaut Barry "Butch" Wilmore, setting up the Rodent Research-1 hardware in the Microgravity Science Glovebox aboard the ISS.

THE VEGETABLE PRODUCTION SYSTEM

Plants and space have a long history. As far back as 1946, NASA launched seeds aboard a V2 rocket commandeered from Nazi Germany after World War II. Five hundred tree seeds flew around the Moon on Apollo 14 in 1971 and were planted on Earth upon their return, with no notable differences between the resulting plants and normal trees.

More recently, astronauts have been cultivating plants aboard the ISS, using the Vegetable Production System known colloquially as Veggie. Measuring 21 by 16 inches (53 x 40 cm) and weighing 16 pounds (7.2 kg), it was installed in 2013 by NASA astronauts Steve Swanson and Rick Mastracchi. Light is supplied by green, red, and blue LEDs, and plants grow on pillows connected to a water reservoir. The pillows are laced with fertilizer to help their growth. Without gravity to tell them which way is up, the roots of space plants tend to grow in all directions.

Astronauts successfully grew red romaine lettuce and sampled the fruits of their labor in 2015. NASA astronaut Scott Kelly said it tasted "like arugula" (aptly called "rocket" in some other countries, such as the UK). In the future, cultivating food in space will be crucial, particularly if we're to send humans to Mars (see pp. 168–171).

v NASA astronauts Kjell Lindgren (left) and Scott Kelly (right) snack on space-grown red romaine lettuce.

EFFECTS OF MICROGRAVITY

YOU SIT IN AN ARMCHAIR IN A NURSING HOME, WITH YOUR GRANDCHILDREN BY YOUR SIDE AND GREAT-GRANDCHILDREN CRAWLING AT YOUR FEET.

The whole room is gripped as you recount the memories of your six-month stay in space. You tell them what Earth looked like during your spacewalk and what it felt like to crash back to gravity after half a year away from the planet.

Now, in your nineties, your bones are starting to give out. You have osteoporosis, a condition that's weakened your skeleton due to your advanced age. Yet, you're already an old pro in dealing with degrading bones. As an astronaut, you've had to cope with it before.

A skeleton doesn't have to work in weightlessness to hold your frame up against gravity. We often think of bones as solid, permanent structures, but they are in fact living tissue. Use them or lose them. Left unaddressed, an astronaut on the International Space Station (ISS) would lose 1–2 percent of their bone density for every single

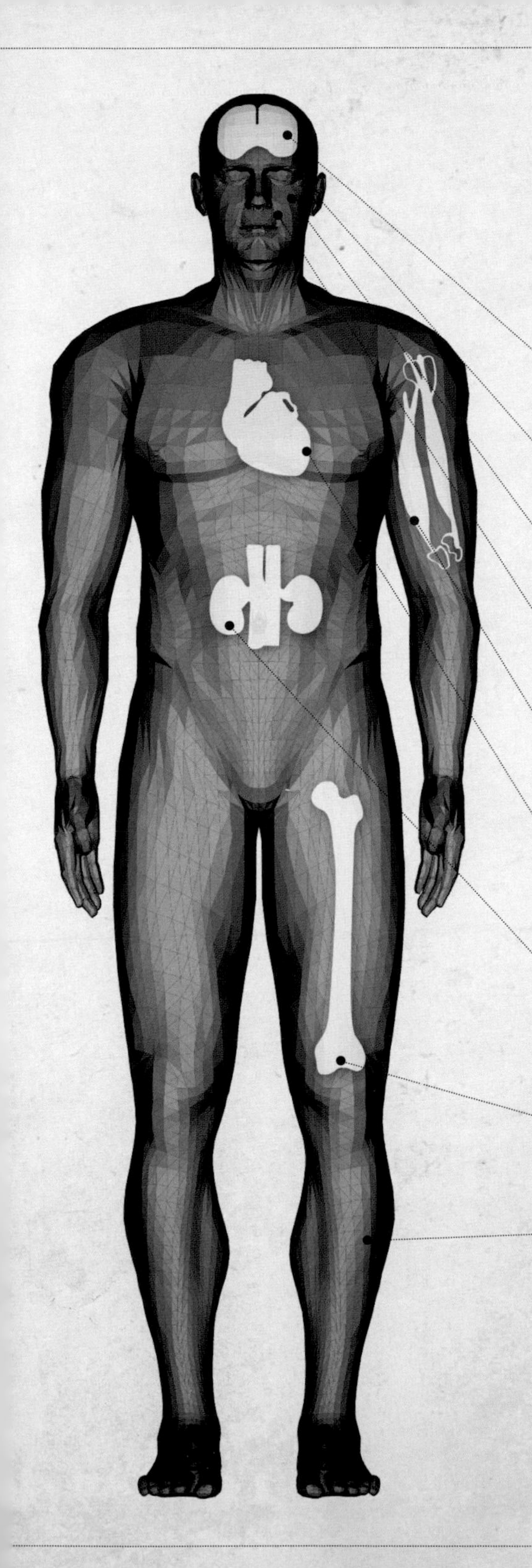

THE EFFECTS OF WEIGHTLESSNESS

Exposure to high-energy cosmic radiation increases risk of neurodegenerative disease.

Eye abnormalities may arise due to UV exposure and increased intracranial pressure.

Due to microgravity, distribution of fluids to the upper body result in a "puffy face" appearance.

Nasal congestion due to fluid redistribution causes anosmia (loss of smell) and diminished taste.

Musculoskeletal adaptation to microgravity leads to loss of muscle mass and bone density.

Heart stroke volume decreases as cardiovascular system adapts to microgravity. Red blood cell number also decreases.

Blood plasma volume is reduced by increased kidney output. Elevated calcium secretion results in increased risk of kidney stones.

Stresses of space flight, including ionizing radiation, result in compromised immune system function.

Fluid redistibution from legs to upper body results in 10–30 percent decreased leg circumference.

Weightlessness has many ways of reeking havoc on the human body from head to toe.

< Canadian astronaut Chris Hadfield juggles tomatoes in microgravity aboard the ISS.

CHICKEN LEGS AND PUFFY FACE

Without normal Earth gravity, there is nothing to pull the fluids in your body away from your heart and down into your lower regions. This leads to a buildup of liquid in the upper body and a lack of it in your legs. Astronauts refer to these two conditions as "puffy face" and "chicken legs." Space travelers often report having a stuffy head, as if they have a blocked nose and sinuses. This can affect their vision as well as their sense of smell and taste.

Your body's natural reaction to too much blood in its top half is to assume that there is an excess of blood in your entire body. Consequently, it tries to shed liquid by making you need to urinate. Most astronauts have the urge to pee soon after entering weightlessness. This decreases your levels of plasma and red blood cells, which isn't a problem until you return to Earth, when might experience dizziness or even faint.

In addition, without the need to pump blood back up the body against gravity, the heart weakens and your blood pressure drops. One study suggests astronauts' hearts get more spherical during stays in space.

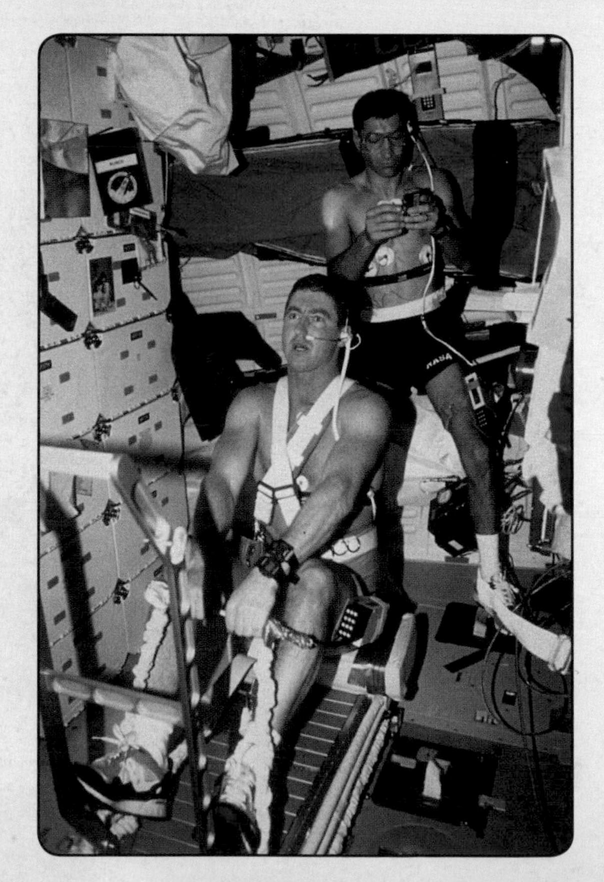

∧ Astronaut Tom Henricks rowing on a modified treadmill for biomedical tests and exercise with colleague Mario Runco Jr.

month they spend in orbit. A six-month stay and your pelvis degrades by 10 percent. That's about the same as a decade's worth of natural aging. You'll have elevated levels of calcium in your body as your bones slowly begin to dissolve back into your bloodstream. This significantly raises the risk of kidney stones—a minor issue on Earth, but a potentially mission-aborting problem in space.

These well-known conditions have focused the attention of space medics to look for solutions. Astronauts spend fifteen hours a week working through a highly regimented exercise regime (see pp. 72–75) to keep their bones and muscles strong. Even then, it can take three to four years for your skeleton to recover once you return to Earth. In 2015, biologists trialed a method of expelling kidney stones using ultrasound instead of the conventional surgical approach. Japanese researchers have experimented with giving space travelers the osteoporosis drug bisphosphonate a week before launch. The result was less calcium in their urine.

Muscle atrophy is also a big deal for the same reason that bones decay. Researchers took biopsies from the calf muscles of nine astronauts before and after a 180-day trip into space. Their capacity for physical work dropped by more than 40 percent. That's the same as turning the astronauts into the equivalent of eighty year olds in just six months. Muscles and bones are often linked together, magnifying the issues. An astronaut's spine starts to straighten without a downward pull of gravity. That's why two-thirds of returning astronauts report back pain problems. They are also four times more likely to experience a herniated (slipped) disk.

Scientists face a constant battle between nature and nurture when running experiments with human subjects. Are the results they see due more to our genes (nature) or our interactions with the environment around us (nurture)? One way to get answer these issues is to study identical twins, two siblings with the same genetic makeup. Any differences between them should be largely due to nurture.

NASA took advantage of a unique opportunity to perform the ultimate space twin study when identical brothers Mark and Scott Kelly were both in the astronaut pool. Scott, who is six minutes younger, became the first American to spend a year (340 days) on board the ISS between 2015 and 2016. Mark stayed on the ground going through a similar routine. Both men had samples taken before, during, and after Scott's mission.

The preliminary results show that space flight changed the way thousands of Scott's genes were turned on and off, and those changes can last for weeks after returning home. Scott's telomeres also got longer—telomeres are the caps at the ends of chromosomes that stop them from deteriorating. They returned to normal after he landed, but the findings could be used to fight aging.

v NASA performed the ultimate twins study with brothers Mark Kelly (left) and Scott Kelly (right).

EXERCISE

KEEPING FIT IN SPACE IS CRUCIAL, SO MUCH SO THAT ASTRONAUTS ON THE INTERNATIONAL SPACE STATION (ISS) HAVE TO EXERCISE TWO-AND-A-HALF HOURS A DAY, SIX DAYS A WEEK.

It's the best way to combat the crippling effects of muscle and bone loss caused by weightlessness (see pp. 68–71). NASA astronaut Clayton C. Anderson found that exercise helped his mental state as much as his physical well-being. A fitness routine is an indispensable part of any long-duration space mission, including any future human missions to Mars (see pp. 148–171).

Half the time is typically spent lifting weights, and the rest on the treadmill or an exercise bike. But how do you lift weights when nothing actually weighs anything? Free weights won't work, so scientists at places such as NASA's Countermeasures Training Facility have designed clever alternatives. To pump iron, astronauts use the Advanced Resistive Exercise Device (ARED), delivered to the ISS by the Space Shuttle Endeavour in 2008. The ARED is made from two cylinders with internal pistons. You pull against the pressure created by the vacuum inside. Astronauts place a bar across their shoulders behind their heads and work through a series of squats, heel lifts, and dead lifts.

Without gravity, a space treadmill doesn't need to be on the floor—it works just as well on the wall or the ceiling. You have to run in a harness, however, to keep you from floating away. Astronauts have even used the treadmill to run marathons in space (see box, p. 75). One of the treadmills that has been used on the ISS is known as COLBERT. This is a "backronym" for US talk show host Stephen Colbert. NASA ran a public competition to name one of the space station's

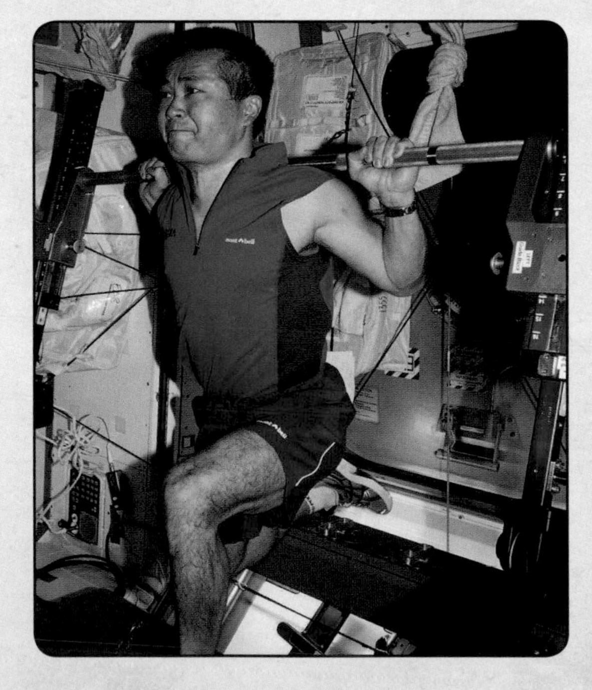

∧ Japanese astronaut Koichi Wakata using the Advanced Resistive Exercise Device (ARED) on the ISS.

> Canadian astronaut Bob Thirsk installing the COLBERT treadmill on the ISS.

THE MUSCLE ATROPHY RESEARCH AND EXERCISE SYSTEM (MARES)

Our understanding of the effect of microgravity on astronauts' muscles, along with how to keep muscle atrophy at bay, is still in its infancy. That knowledge needs to be developed further if we want to send the first humans to Mars and make safe space tourism an everyday occurrence. The astronauts of today are opening the door for all of us.

As part of these efforts, the Muscle Atrophy Research and Exercise System (MARES) was installed on the ISS in 2015. European Space Agency astronaut Samantha Cristoforetti set up and calibrated the 440-pound (200-kg) device, and fellow ESA astronaut Andreas Mogensen was the first subject. It is installed in the ISS's European Columbus module and is so big that it has to be stowed away when not being used.

While they sit in a built-in chair, MARES monitors astronauts' muscles as they exercise, providing a comprehensive overview of muscle speed and force at key joints, such as elbows and knees. Astronauts move their joints to follow a dot on a screen as the device's motor generates a counterforce. The motor was designed from scratch, because no existing motor could cope with the unique challenges of the weightless environment in orbit.

> British astronaut Tim Peake using the European Muscle Atrophy Research and Exercise System (MARES) on the ISS.

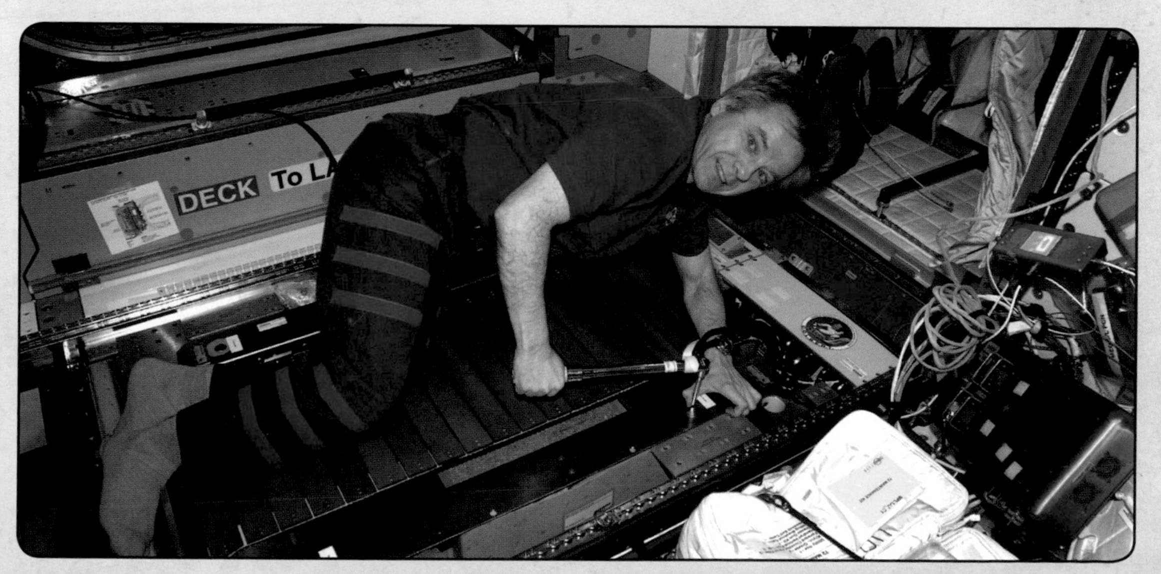

nodes, and Colbert won the vote after enlisting the help of his viewers. NASA exercised its discretion and named the running apparatus the Combined Operational Load Bearing External Resistance Treadmill instead. Astronauts can use it to run at speeds of up to 12 miles (20 km) per hour.

The exercise bike—or Cycle Ergometer with Vibration Isolation System (CEVIS)—is used for daily exercise, and also for prelanding fitness checks. Astronauts wear a heart rate monitor, and the readout is shown on a control panel along with other information, including their pedaling speed (which can reach 120 revolutions per minute). Often they watch movies while working out to pass the time. The data is then downloaded onto a flash drive, before astronauts upload it to the Station Support Computer (SSC) for analysis by scientists on the ground. The results show that astronauts' exercising heart rates are higher at the start of the mission, but fall more in line with their preflight levels as they acclimatize to weightlessness.

v NASA astronaut Catherine "Cady" Coleman using the Cycle Ergometer with Vibration Isolation System (CEVIS) in the Destiny laboratory on the ISS.

RUNNING A MARATHON IN SPACE

While thousands of nervous runners lined up to start the 2016 London Marathon, British astronaut Tim Peake was 273 miles (400 km) above the Pacific Ocean. That didn't stop him from joining in. Running on the ISS's treadmill, the forty-four year old completed the 26.2-mile (42-km) distance in an impressive three hours, thirty-five minutes, and twenty-one seconds (seventeen minutes slower than when he ran the race for real in 1999). During the race, he orbited Earth almost two-and-a-half times, traveling a distance close to 62,140 miles (100,000 km). His efforts saw him become the second person to run a marathon in space, after NASA astronaut Sunita Williams, who completed the Boston Marathon in 2007. Peake's time was faster, setting a Guinness World Record.

To allow for him to run in a weightless environment, Peake was strapped into a harness that gave him 70 percent of his body weight on Earth. He had to fight through the pain barrier when the harness dug into his flesh, leaving him with nasty abrasions and sores on his shoulders and around his waist. He had worked with an exercise specialist for two years to adjust his running style to minimize the chance of further injuries.

v NASA astronaut Sunita Williams exercising on the COLBERT treadmill on board the ISS.

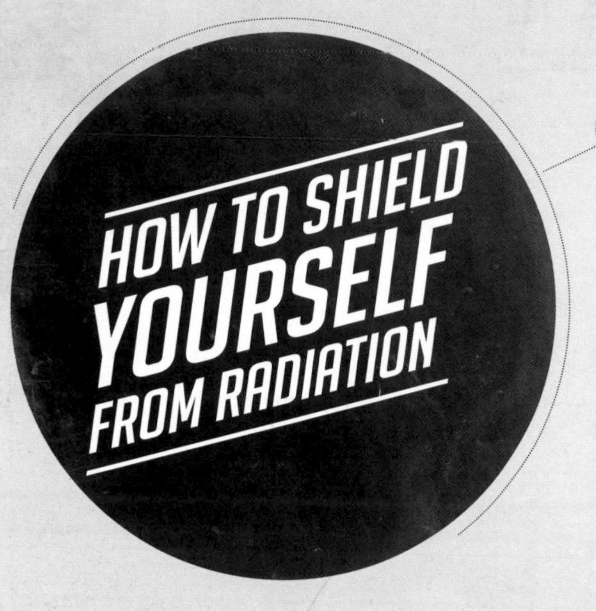

HOW TO SHIELD YOURSELF FROM RADIATION

WE HUMANS HAVE EVOLVED INSIDE THE BLISSFUL COCOON OF EARTH'S MAGNETIC FIELD. THIS BLANKET AGAINST THE HARSH REALITY OF A RADIATION-FILLED COSMOS KEEPS OUR FRAGILE BIOLOGY SHIELDED FROM THE WORST THE UNIVERSE HAS TO THROW AT US.

For those brave enough to venture into space, the protection from the Earth's invisible magnetic force field is diminished. Travel beyond low Earth orbit (LEO), perhaps to the Moon or Mars, and it disappears entirely. The 24 astronauts who left LEO as part of the Apollo missions were just lucky they didn't encounter bigger problems. Even astronauts on the International Space Station (ISS) must take heed, because they sit higher in the planet's magnetic field and so are offered less shelter.

Too much radiation can wreak havoc with the human body (see box, opposite), so trying to minimize astronauts' exposure is key to their survival. One of the biggest culprits for producing this life-endangering energy is the Sun—the very object normally associated with supporting life. The Sun looks steadfast and stable, but that could not be farther from the truth. Instead, it is a seething, churning, undulating sea of roiling plasma. Its own magnetic field twists and contorts as it rotates, storing up energy much the same way as winding up a rubber band. Eventually, the energy can no longer be contained and the Sun erupts. These magnetic mood swings—known as coronal mass ejections (CMEs)—can involve more than a billion tons of

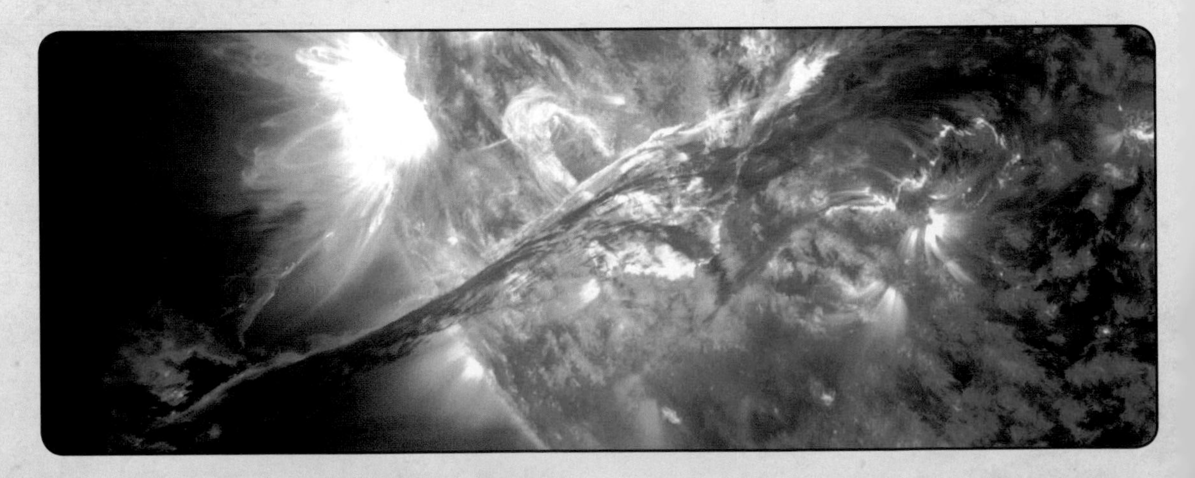

∧ Sunrise from the ISS Zvezda module, one of the safest places to shelter from solar activity.

↲ A coronal mass ejection, traveling at more than 900 miles (1,500 km) per second, taken by the Solar Dynamics Observatory.

solar material rocketing out into space at a million miles an hour. Such eruptions are often accompanied by a searing flash of visible light and X-rays—a solar flare.

Both CMEs and solar flares can pose a real danger to astronauts, so mission control constantly monitors the Sun for any sign of increased solar activity. Fortunately, the threat can be avoided as long as certain precautions are taken. For example, any planned spacewalks would be canceled, because the hull of the ISS itself acts like a skin, shielding the space travelers inside from much of the danger. For extra peace of mind, during storms, astronauts are moved to the parts of the ISS that offer the greatest protection. These include the aft ends of the US Destiny laboratory and the Russian Zvezda service module, which is built heavily enough to reduce the crew's radiation exposure by 60 percent. Even then, they might get a normal week's worth of radiation in just twelve hours. This run-for-cover exercise is exactly what happened in both 2000 and 2006. In the latter case, the crew was instructed to sleep in these protected areas in case an incoming storm worsened during their slumber.

WHAT EFFECT DOES RADIATION HAVE ON THE BODY?

There are two main ways that space radiation can damage your body, and both affect your DNA. First, the water in your body can absorb the radiation, causing a reaction that creates a particle known as a free radical. These free radicals can then interact with your DNA, breaking the bonds that form its double helix structure. Alternatively, the radiation itself can damage the DNA directly, causing the cell in which it is contained to change or die.

If the radiation dose is low enough, your body can often repair the damage. Too high, however, and you can expect radiation sickness or even death. Altered (mutated) cells can go on to copy themselves incorrectly, and they may eventually become cancerous. According to NASA, when it comes to radiation, future onset of cancer is the biggest concern for astronauts. During a six-month stay on the International Space Station, astronauts are exposed to at least 50 millisieverts (mSv) of radiation. Heightened cancer risk is a known effect of radiation above this level.

Solar flares are grouped into five categories from weakest to strongest: A, B, C, M, and X. The first four groups are subdivided into ten categories by giving them numbers 0–9: for example, a M5 flare. X is the last and most powerful group, but there is no upper limit. In 2003, one erupting flare was so powerful that detectors stopped working after measuring it as an X28. It has since been estimated as reaching the X45 level. Fortunately, that one was aimed away from Earth. Had it not been, our atmosphere would have been bombarded by X-ray radiation equivalent to that of 5,000 Suns.

It is the M- and X-class flares that are of particular concern to mission controllers and orbiting astronauts, with telescopes, such as NASA's Solar Dynamics Observatory, keeping our star under constant surveillance. Had an X-class solar flare erupted during any of the Apollo missions to the Moon, the astronauts would have been exposed to an estimated 500 mSv of radiation—enough to make them extremely ill and 10 times more than the raised cancer risk limit.

X-RAY FLUX NEAR EARTH, MEASURED IN WATTS PER SQUARE FOOT

LESS THAN 0.000001 WATTS PER SQUARE FOOT → GREATER THAN 0.0001 WATTS PER SQUARE FOOT

(A) (B) (C) (M) (X)

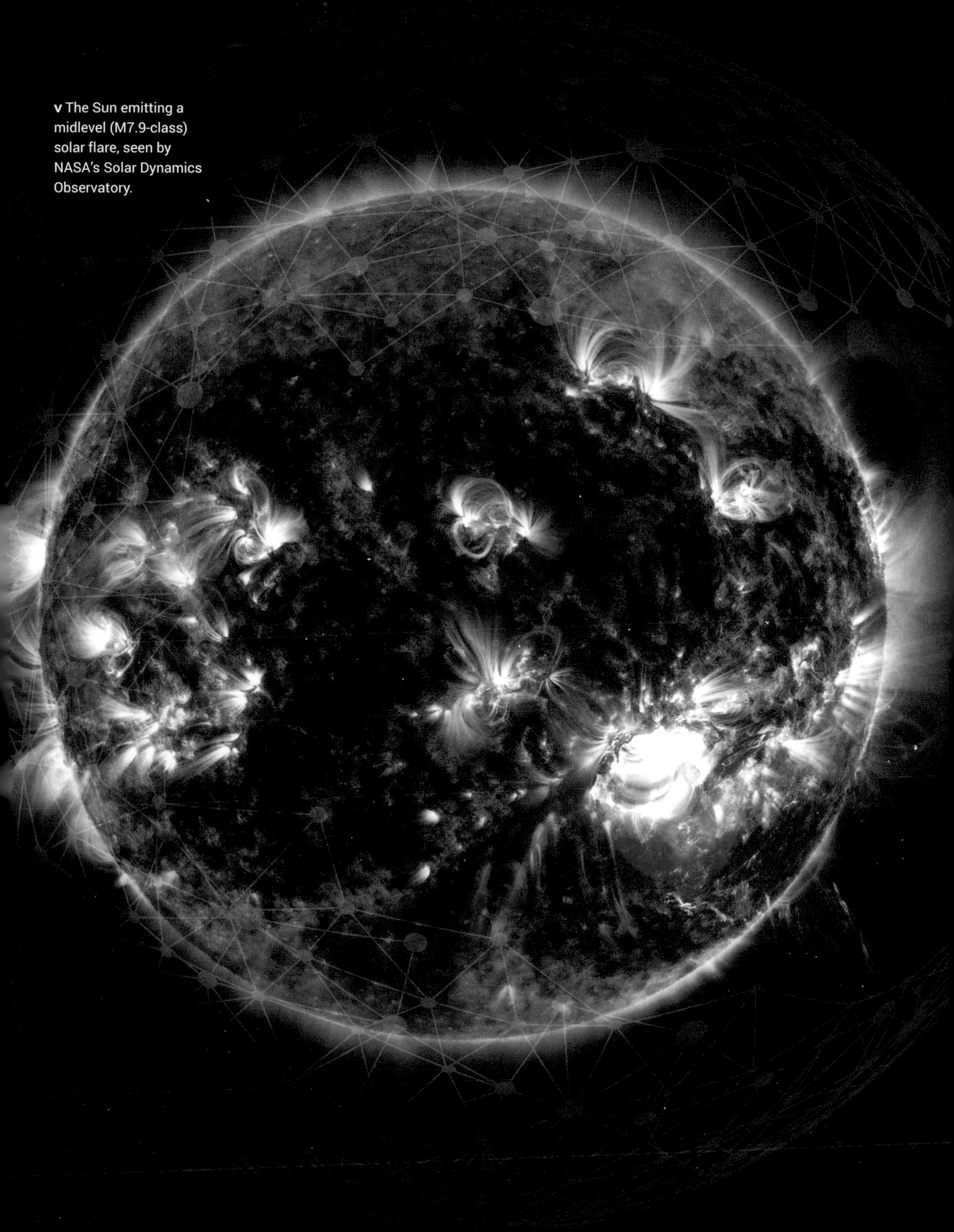

v The Sun emitting a midlevel (M7.9-class) solar flare, seen by NASA's Solar Dynamics Observatory.

SPACEWALKS

AFTER YEARS OF TRAINING AND HUNDREDS OF CHECKS, PROCEDURES, AND PREPARATIONS, YOU ARE READY FOR THE MOST INCREDIBLE EXPERIENCE OF YOUR LIFE—IT'S THE DAY YOU GET TO WALK IN SPACE.

Officially, these jaunts outside an orbiting spacecraft are called extra-vehicular activities (EVAs). To everyone else they are known as spacewalks. The hatch opens, and for the next six-and-a-half hours it is just you, your colleague, and the magnificence of Earth set against a starry background.

Spacewalks normally happen because something needs repairing or a new part needs installing. You pull yourself along the external hull of the space station, deftly maneuvering into the right position. NASA astronaut Doug "Wheels" Wheelock says the word "spacewalk" is a misnomer, describing it instead as "a space ballet on fingertips." He should know—he's done six of them. A few astronauts have even performed untethered spacewalks (see box, p. 83).

Astronauts can drink through a straw and go to the toilet using the wonderfully titled Maximum Absorbency Garment (MAG)—an adult diaper. They deliberately eat light meals ahead of an EVA—a protein bar, granola, and fruit are common choices.

One of the biggest challenges is contending with the rapidly changing temperatures; day turns to night multiple times during a spacewalk. Work is often arduous, and things don't always go to plan. Don't expect to be able to vent your frustration with swear words either—your microphone is always live for communication with your fellow spacewalkers and mission controllers.

∧ NASA astronaut Stephen Robinson riding the robotic arm during STS-114, doing a first in-flight repair of a Space Shuttle.

ALEXEI LEONOV'S MAIDEN SPACEWALK

On March 18, 1965, Russian cosmonaut Alexei Leonov made history when he tentatively ventured outside his Voskhod 2 capsule. He became the first astronaut to perform an extra-vehicular activity (EVA) or spacewalk. He later said: "You just can't comprehend it. Only out there can you feel the greatness—the huge size of all that surrounds us." His landmark feat only lasted 12 minutes and nine seconds, but it almost ended in disaster.

Leonov's space suit ballooned in the vacuum in space, rendering him too big to fit back through the hatch. Five minutes longer and night would fall, plunging him into darkness. Thinking quickly, he vented air through a valve in the lining of his suit, but that meant losing oxygen and risking decompression sickness. Panicked, Leonov entered the airlock head first instead of feet first. He was sweating so much he couldn't see anything. Eventually, he made it safely inside.

It wasn't the last of Leonov's troubles. Oxygen levels rose to a level that posed a real fire risk on reentry. They were stabilized, but then the reentry rockets failed to fire and the crew had to trigger them manually, shunting the landing way off target into a Siberian region full of bears and wolves.

v Russian cosmonaut Alexei Leonov became the first man to walk in space during the Voskhod 2 mission in 1965.

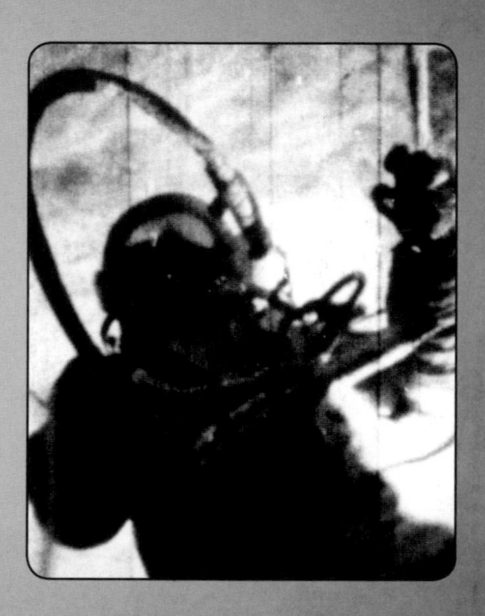

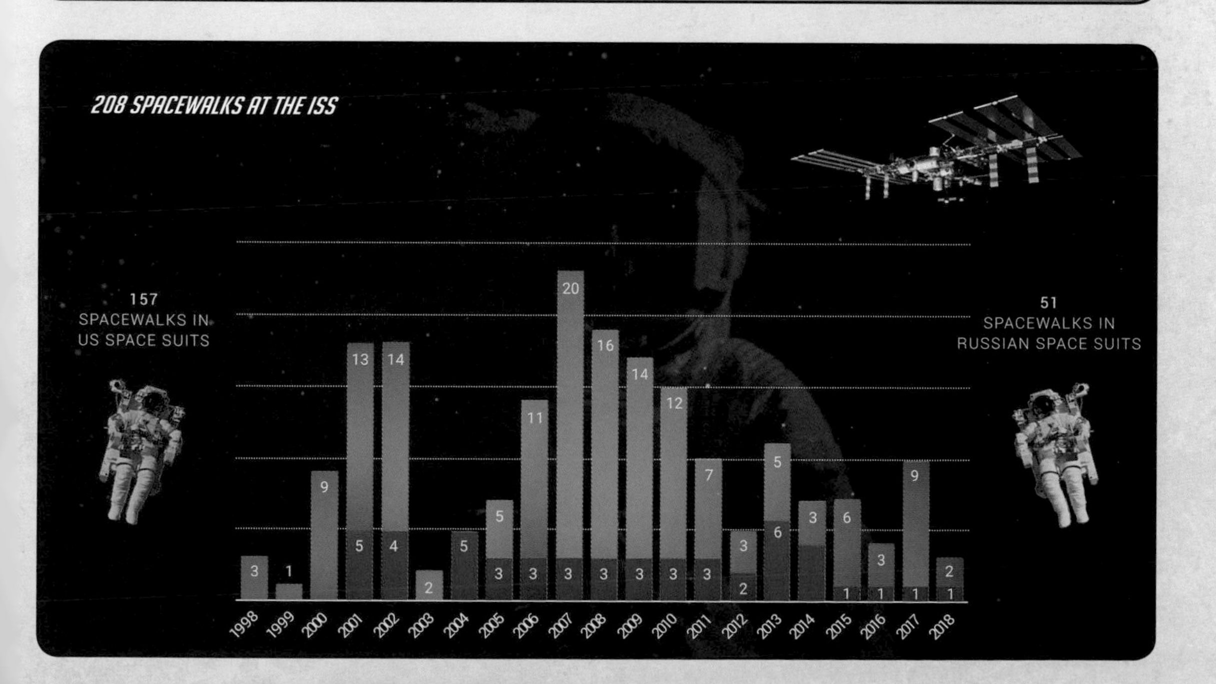

Spacewalks might be exhausting, but the experience is unrivaled. Perhaps Wheelock describes it best. "Earth is this living, breathing ball of life in this vast empty sea and it's just raging with light and life and motion and color," he says. "It's just quite amazing." Former International Space Station (ISS) Commander Chris Hadfield is equally animated. "It is like coming around a corner and seeing the most magnificent sunset of your life, from one horizon to the other, where it looks like the whole sky is on fire and there are all those colors," he says. "The Sun's rays look like some great painting up over your head. You just want to open your eyes wide and try to look around at the image, and just try and soak it up. It's like that all the time."

However, it's not all spectacular views. Sometimes things go badly wrong, just as with Alexei Leonov's first spacewalk (see box, p. 81). More recently, ESA astronaut Luca Parmitano had a lucky escape when his helmet began filling with about 6 cups (1.5 liters) of water during an EVA in 2013. He nearly drowned. The weightless liquid filled his eyes, ears, and nose, so he struggled to see or tell anyone what was happening. To make matters worse, the Sun had just set. Fortunately, he managed to make it back to the airlock in time.

< NASA astronaut Robert Curbeam (left) and ESA astronaut Christer Fuglesang (right) during a spacewalk to construct the ISS in 2006.

∧ British astronaut Tim Peake during a spacewalk to repair part of the ISS's solar panels in December 2015.

UNTETHERED SPACEWALKS

Leaving the relative safety of a spaceship takes some chutzpah, but wandering in space free from an umbilical cord tethering you to the mothership takes things to another level. On February 7, 1984, during a mission on the Space Shuttle Challenger, NASA astronaut Bruce McCandless became the first person to perform an untethered spacewalk. He was strapped to a Manned Maneuvering Unit (MMU) instead—essentially a jet pack—allowing for him to move around at will. McCandless's colleague, Robert L. Stewart, followed soon afterward.

The unit worked by firing nitrogen out of twenty-four nozzles operated by the astronaut using hand controllers at the end of two fixed arms. The left controllers covered the nozzles for up-down, left-right, back-front motion, whereas the right hand was for rotation. The plan was for astronauts to approach faulty satellites and drag them into the shuttle for docking and repair. However, that's easier said than done. A few months after McCandless's mission, Challenger carried James van Hoften and George Nelson into orbit, where they also used the MMU to try and grab the faulty SolarMax satellite. They were thwarted by a part that wasn't on the satellite's blueprint and had to use the shuttle's robotic arm instead.

> NASA astronaut Bruce McCandless during his untethered spacewalk on February 7, 1984.

SLEEP

FRESHLY CHANGED INTO YOUR STANDARD-ISSUE, FULL-LENGTH, GREEN PAJAMAS, YOU FLOAT DOWN THE CAPSULE TO REACH YOUR SLEEPING STATION IN NODE 2 OF THE INTERNATIONAL SPACE STATION (ISS).

You open the concertina doors of your sleeping station and move inside. On the walls are pictures of your family. Good-luck charms and mascots hang inside protective nets. They make your private little soundproof box feel more like home. Closing the doors shuts you away from your fellow crew mates—an orbital oasis of peace and quiet (almost).

Before you go to sleep, you're eager to download the day's photographs of Earth from your camera's memory card to your laptop. Checking your emails, you find a message from Mission Control about tomorrow's activities. To reply, you first have to curl your toes under a bar on the floor—otherwise the pressure of hitting the

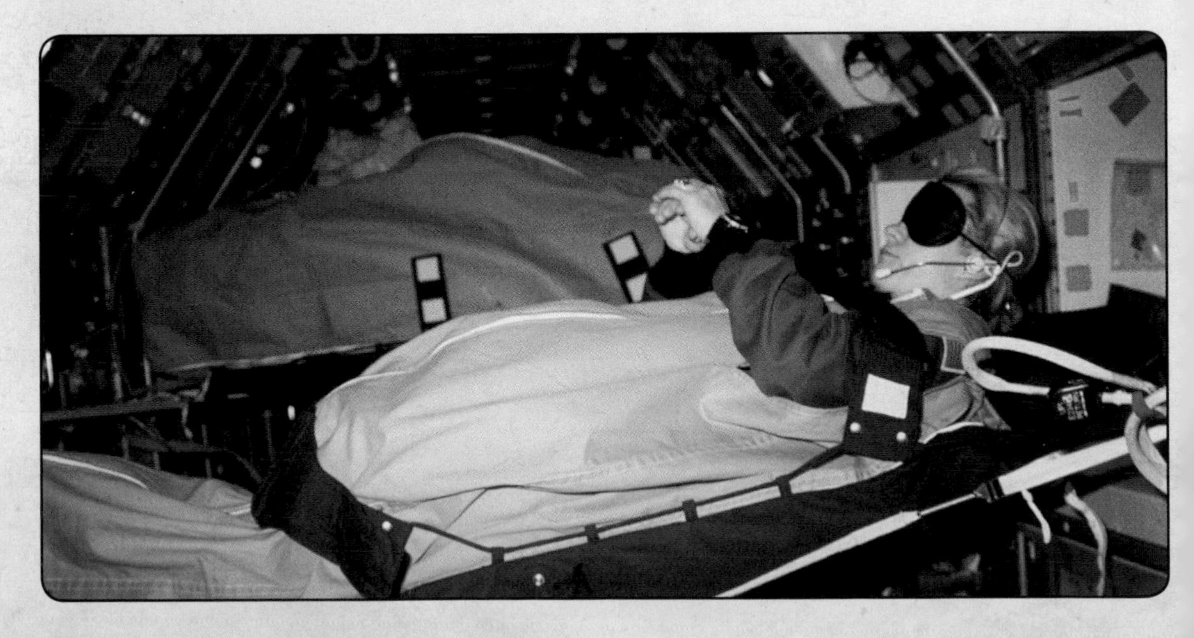

DO ASTRONAUTS HAVE SPACE ALARM CLOCKS?

If astronauts are supposed to sleep for around eight hours, even when it's light outside, how do they know when to emerge from their sleeping bags? Until the end of the Space Shuttle era, NASA woke its astronauts using music—a tradition stretching back to 1965, when the *Gemini VI* crew were played *Hello, Dolly!* by Jack Jones. Today, astronauts on the ISS are still occasionally roused with a song.

The person in charge of the music is CAPCOM, the capsule communicator in Mission Control (see pp. 40–43). Songs with a space theme are common, including Elton John's "Rocket Man" and David Bowie's "Space Oddity." Michael Stipe, lead singer of REM, once sang an a capella version of their hit "Man on the Moon." Clips from TV shows or movies are sometimes used, too. In 1991, the crew of the Space Shuttle Atlantis woke to the voice of Patrick Stewart reciting lines from *Star Trek: Next Generation*. James Earl Jones, who provided the voice of Darth Vader in the original *Star Wars* trilogy, also took on the task. The last song played on the final Space Shuttle mission was Kate Smith's rendition of "God Bless America."

> Gemini VI astronauts Wally Schirra (seated) and Thomas Stafford were the first to be woken up with a song back in 1965.

keyboard with your fingers would see you rise toward the ceiling. Some astronauts attach a cord to the bar and run it around their waist to anchor them in place while working. Don't expect a fast Internet; it is apparently like the glacial old days of dial-up.

< NASA astronaut Margaret Seddon sleeps with a blindfold during the nine-day Space Shuttle Columbia mission in June 1991.

You're ready for bed now, but there are no beds on the ISS. A sleeping bag is attached to one of the walls of your cabin with bungee cords instead. With no need to support your body and head against gravity, you sleep vertically in space, not horizontally. You poke your arms through the holes, and as you relax, your hands float out in front of you, making you look a little like a marauding zombie. Some astronauts find it difficult to sleep like this, so they fold their arms or tuck them inside the sleeping bag. You

Orbiting the planet every ninety-two minutes has its advantages, including sweeping views of the turning continents below and sixteen spectacular sunrises and sunsets every day. Yet, constantly changing light levels—switching from day to night every forty-five minutes—takes its toll on your sleep patterns.

Our earthly sleep-wake cycle is controlled by a twenty-four-hour internal clock known as our circadian rhythm. Light entering the eyes strikes a tiny region in the hypothalamus of the brain, called the suprachiasmatic nucleus. When light levels drop, it triggers the pineal gland to release the hormone melatonin, making us feel sleepy. This natural cycle is severely disrupted in space.

In January 2016, scientists published the results of a study analyzing the sleep patterns of twenty-one astronauts over a cumulative total of more than three thousand days in orbit. They found that the average sleep duration was just six hours, when it was scheduled to be eight and a half. Three-quarters of astronauts reported using sleep medication, such as zolpidem and zaleplon. That is changing, however, with astronauts undergoing cognitive behavioral therapy (CBT) as part of their training in order to learn how to tackle insomnia without medications.

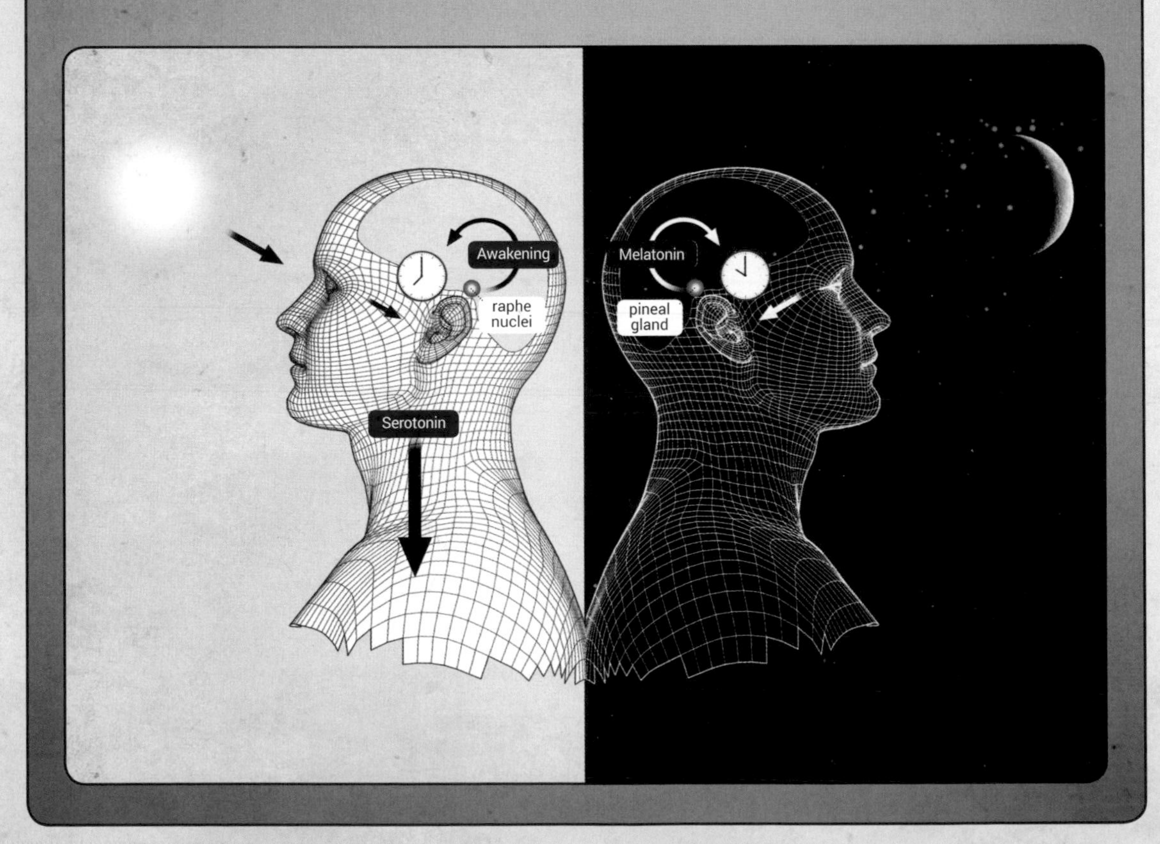

v Our sleep-wake cycle is governed by changing light levels. In space, this cycle can be disrupted.

can shorten the cords so you're more tightly bound to the wall. Others enjoy the freedom of floating around the cabin during their slumber, although you might just bump into things and startle yourself awake. Being trussed up also means it takes a longer to free yourself if you need to make a trip the bathroom in the night.

You lower a blindfold to keep it dark, otherwise you'd be awoken every forty-five minutes by yet another sunrise (see box, left). A wearable device on your wrist keeps track of your sleep patterns, feeding data back to Mission Control for analysis. Changing light levels are just one reason why it's not always easy to sleep in orbit. Astronauts frequently report being too hot or too cold, and the ISS is a noisy environment, with the constant industrial hum of heavy machinery. So a lot of astronauts wear earplugs to help them sleep, as long as they can still hear the emergency and wake-up alarms (see box, p. 85).

< ESA astronaut Samantha Cristoforetti inside her sleeping bag during her stay on the ISS.

v Japanese astronaut Koichi Wakata in his sleeping bag attached to the racks in the Kibo laboratory of the ISS.

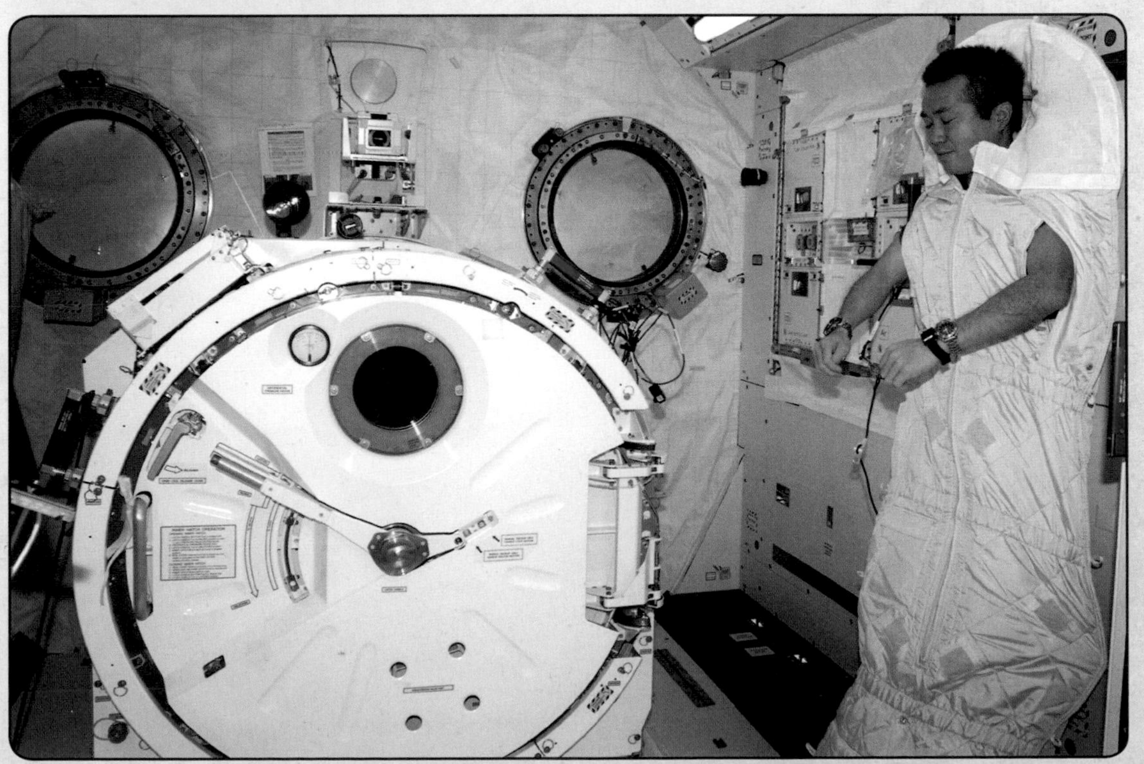

WASHING AND CLEANING

IF YOU THINK YOU CAN ESCAPE DOMESTIC DRUDGERY JUST BECAUSE YOU'RE AN ASTRONAUT, THINK AGAIN.

Saturday is cleaning day aboard the International Space Station (ISS), and all crew members have to chip in to keep the place looking spick and span. Duties include cleaning the meal area, changing air purification filters, vacuuming, and wiping down the walls and floor. Two Russian cosmonauts even cleaned the outside of the windows during a spacewalk in 2015. A little tomfoolery goes a long way to lighten the chores, too—cosmonaut Anton Shkaplerov posted a video of himself using the vacuum cleaner to fly through the space station with an arm outstretched like Superman.

Dust, spillages, and debris must be cleaned up promptly, otherwise they could cause health problems for the crew or damage to equipment. The astronauts' cleaning supplies include plastic gloves, liquid detergent, and cloths—not that different really from life on Earth. They even have disposable wipes on board for dealing with stickier spills. Stray liquids are gathered up in a cloth and then left to evaporate naturally, so that the water heads back into the extraction system to be recycled as tomorrow's coffee. For dangerous spillages, such as battery leaks, the crew are well trained in the use of a contaminated clean-up kit. First, they don protective goggles to stop it from getting in their eyes, then a mask to stop it reaching their nose, mouth, and lungs. Special silver shield gloves protect their hands and arms. Once the offending item has been gathered up with a heavy-duty cloth, it is sealed inside several ziplock bags to ensure no further leaks.

Generally, the ISS is kept clean—certainly cleaner than your bathroom at home. In 2015, scientists from NASA's Jet Propulsion laboratory used the latest genetic sequencing techniques to analyze bacteria found on the station. They used samples taken from an air filter that had been in use for forty months, along with two dust bags taken from the ISS's vacuum cleaner. The results

∧ Japanese astronaut Soichi Noguchi using a vacuum cleaner during housekeeping operations in the Kibo laboratory of the ISS.

↗ NASA astronaut Karen Nyberg demonstrates how to wash your hair in space using a prepared pouch.

WHAT'S PERSONAL HYGIENE LIKE IN SPACE?

You cannot shower in space, at least not in the conventional way. In a weightless environment, the water droplets would just float away instead of falling down over your body and into the drain. So the crew of the ISS clean themselves using something similar to a sponge bath. Spray deodorant doesn't work in microgravity either—it's roll-on all the way.

All personal hygiene items are kept in a Russian-designed kit called the Comfort-1M. Inside is a pouch of body soap to which you add a little warm water. Each pouch has to last two weeks. Soapy water is squeezed onto a washcloth that is changed every other day. After cleaning themselves with the washcloth, astronauts pat themselves down with a towel that's replaced every week. To wash their hair, astronauts use specially designed no-rinse shampoo. Hair is cut right next to the vacuum cleaner to keep the trimmed-off hair from floating away. Fingernails are trimmed adjacent to a ventilation grid to suck up the clippings. Astronaut toothbrushes and toothpaste look similar to yours, but, without a sink, most astronauts choose to swallow after brushing instead of spitting into a cloth.

v NASA astronaut Cady Coleman assists ESA astronaut Paolo Nespoli with a haircut in the Kibo laboratory on the ISS.

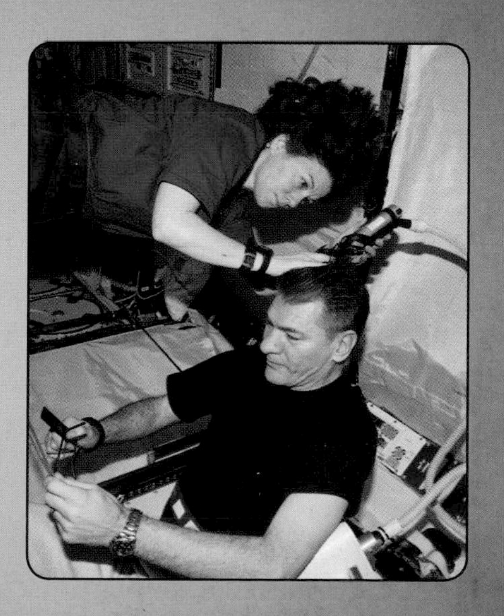

were compared to samples taken from the strictly controlled clean rooms on Earth that NASA uses to build spacecraft.

Skin bacteria called Actinobacteria were more prevalent on the ISS. The vacuum bags contained Staphylococcus, bacteria often responsible for food poisoning. The bags also hosted seventy-five times more bacteria than the filters, showing just how clean the air is up there, despite six people living in such confined quarters for months at a time. The study did not assess the virulence of the bacteria—their ability to actually cause infections or diseases. Living bacteria were even found on the outside of the ISS in 2017 after swabs were taken during a spacewalk. Astrobiologists think these could be bacteria from Earth propelled upward when space dust strikes our atmosphere.

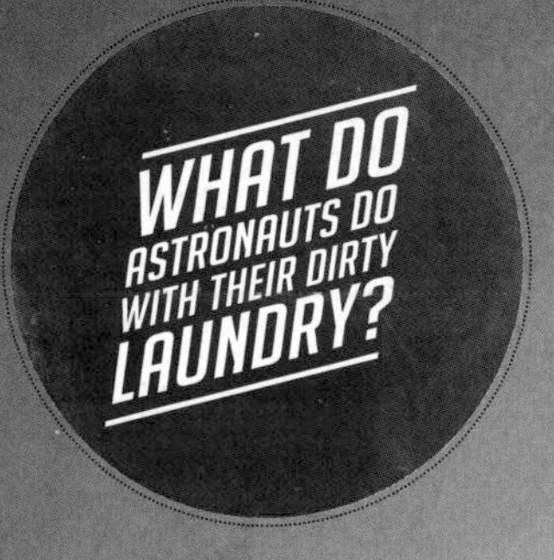

> The uncrewed Progress spacecraft leaving the ISS ahead of burning up in the atmosphere.

Without a washing machine aboard the ISS, space travelers have gotten creative with their used clothes. The default option is to make them last as long as possible. Cosmonauts on the old Russian Mir space station would change their underwear just once a week. The norm on the ISS is more like fresh pants every three or four days. Outer layers of clothing often last a lot longer and can sometimes be worn every day for months.

Space is at a premium, so clothes are discarded once they've reached the point where they can no longer be worn. Usually this means packing them into an empty, uncrewed Russian Progress spacecraft that has just delivered fresh fuel and supplies. The module undocks and is lowered into an orbit that sees it burn up safely over the Pacific. Like their dried poop (see p. 98), astronauts' dirty socks also become shooting stars!

What would happen on a much longer human voyage, say to Mars? Perhaps used clothes could be fed to bacteria. This would create methane, which could be used on board as an additional fuel supply. Scientists working on Mir started designing an experiment to test the idea, but it never flew into orbit.

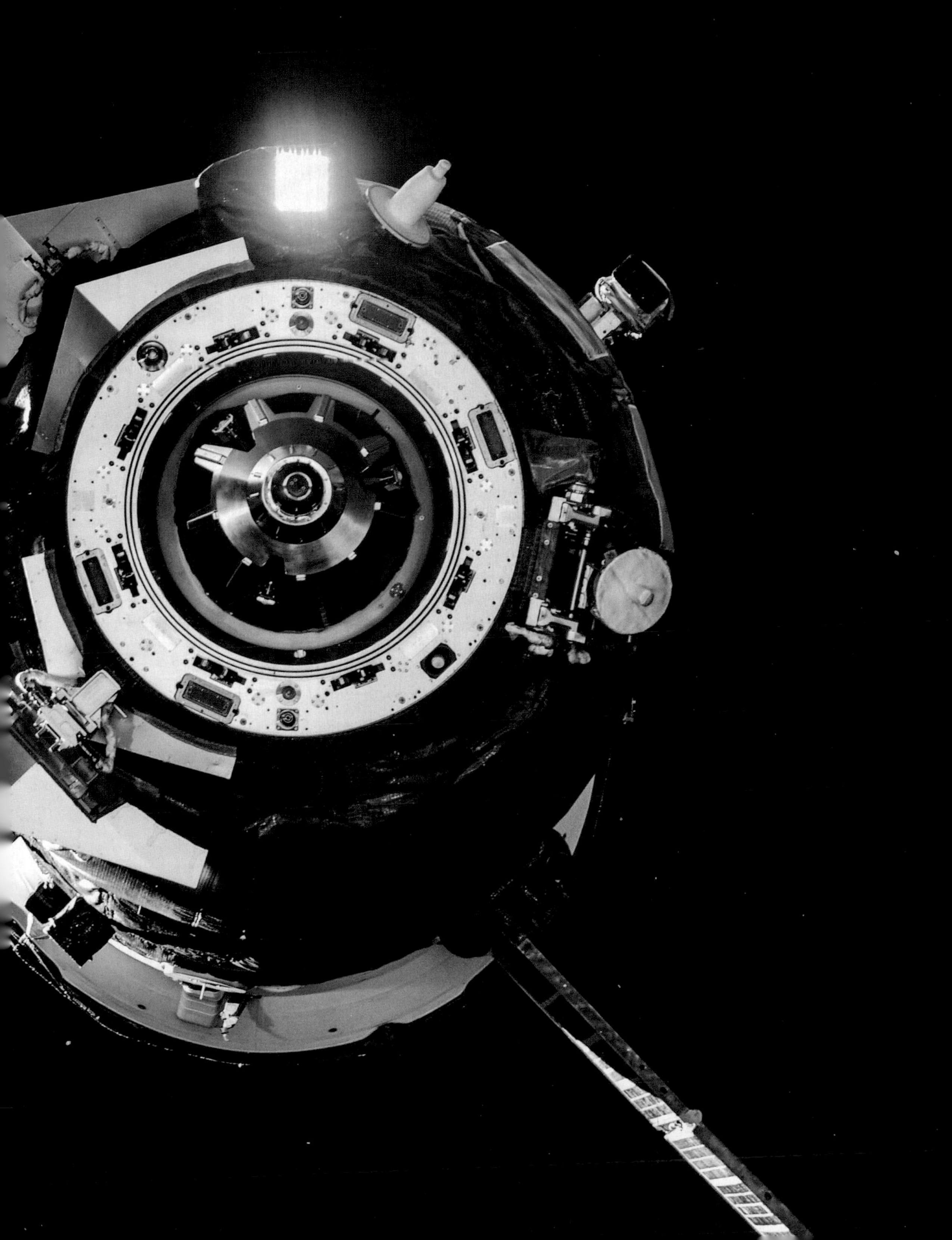

EMERGENCIES

YOU'RE BUSY WITH YOUR WEEKLY CLEANING DUTIES WHEN YOU SEE ONE OF YOUR COLLEAGUES SUDDENLY GRAB FOR THEIR CHEST.

It is clear from the panic on their face that they know they're having a heart attack 250 miles (400 km) up and far from expert medical attention. What do you do?

Fortunately, this kind of emergency has never yet happened on the International Space Station (ISS). However, astronauts undergo forty hours of basic medical training as part of their preparations to go into space. They are schooled in basic first aid, stitching wounds, and extracting teeth. The medical bag on the ISS contains a first aid kit, bags of saline solution, a defibrillator, and an ultrasound machine. So, in the case of a heart attack, the Chief Medical Officer on board (not always a doctor) would probably go for the defibrillator under guidance from the mission physician on the ground. However, you'd hope that constant monitoring of astronauts' vitals would set off alarm bells ahead of time.

Serious medical emergencies could bring the mission to a close, because you'd want to get the patient down to the ground as swiftly as possible. Depending on where you are around the planet at the time, you might be able to undock, perform an emergency reentry, and land in as little as ninety minutes. However, that would mean dealing

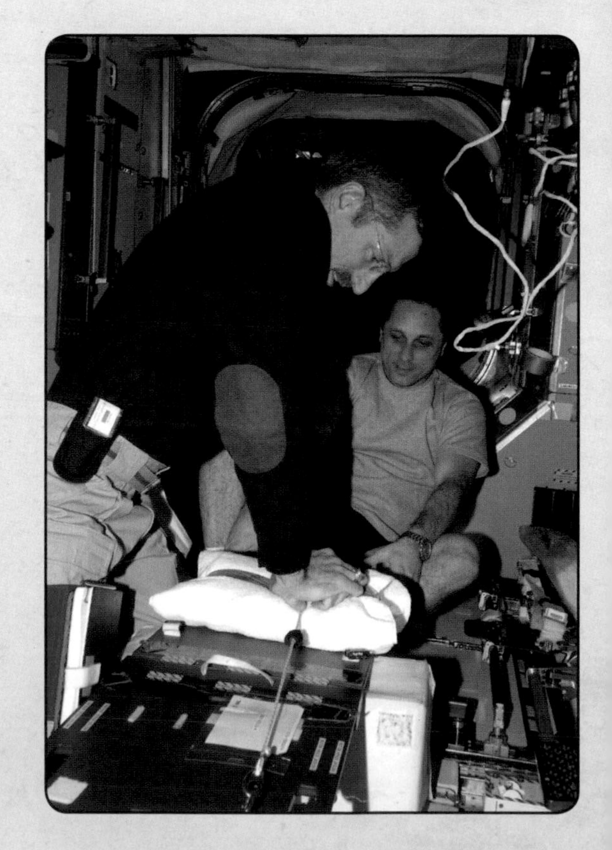

> NASA astronaut Dan Burbank (left) and Russian cosmonaut Anton Shkaplerov (right) during a medical contingency drill on the ISS.

WHAT HAPPENS IF THERE'S A FIRE IN SPACE?

v Flames in space (left) are more spherical than the familar tapered shape they have on Earth (right).

Fire behaves in a different way in free fall than the it does on the ground. On Earth, colder, heavier gas is pulled downward and displaces the gas heated by the flame. That provides the fire with a plentiful supply of new oxygen and gives the flame its familiar tapered shape. Oxygen is drawn to the flames one hundred times more slowly in microgravity. As a result, space fires often consume less oxygen, burn with a lower temperature, and form spherical balls. However, a fire will follow a spacecraft's ventilation system in search for more oxygen.

Between 2008 and 2018, NASA operated two FLame Extinguishment Experiments (FLEX 1 and 2) aboard the ISS. Hundreds of tests showed something remarkable. Drops of fuel can continue burning even after the flame has been extinguished. This means that astronauts need to be sure that a fire is out.

Weightless smoke doesn't rise, so detectors on the ISS are installed in the ventilation system, not the ceiling. Astronauts are trained to don breathing masks as soon as the alarm sounds and turn off the vents to reduce the flow of oxygen. The masks are attached to O_2 canisters, providing 10–15 minutes of oxygen. The crew can then attempt to put out the fire or run for cover to the Soyuz, ready for a swift exit.

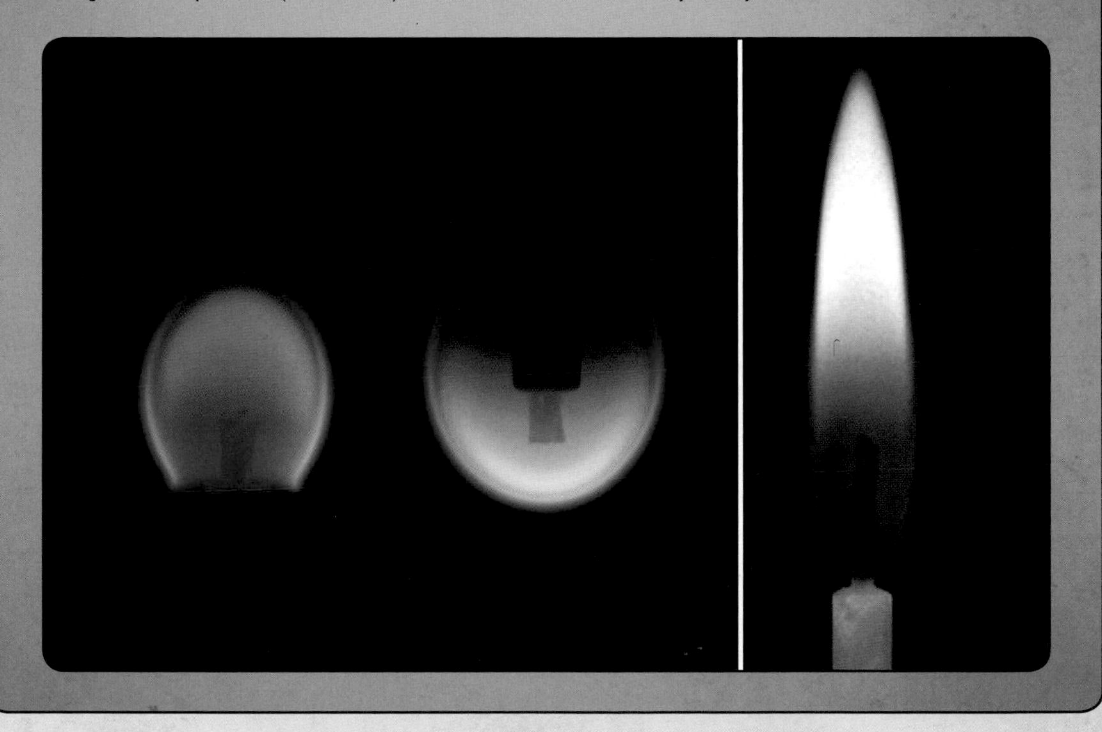

with some high G-forces (see pp. 24–27), and the nature of the injury or accident would determine whether this was feasible. Mission controllers calculate the chances of such emergencies at 1 to 2 percent per astronaut per year.

Apart from medical emergencies, there's a trio of possible incidents that mission controllers fear the most. Astronauts train extensively to deal with all of them. They are known as "The Big Three": fire (see box, above), depressurization (see box, p. 95), and an ammonia leak.

Ammonia runs along pipes on the outside of the ISS to act as a coolant for its power system. If the pipes are breached, the noxious gas can flood the station and suffocate the crew within minutes. In May 2013, astronauts noticed what looked like snowflakes out of the window. These turned out to be ammonia crystals frozen by the coldness of space. A pipe was leaking, but fortunately not into the ISS itself, and the fault was repaired on a spacewalk. The situation seemed even more dire in 2015, when the crew were forced to don protective breathing apparatus and abandon the US end of the ISS, taking shelter in the Russian module and closing the hatch behind them. The ammonia leak alarm had sounded, but thankfully it was a false alarm and they were allowed to return eleven hours later.

v NASA astronaut Shane Kimbrough holds the Robotic External Ammonia Leak Locator aboard the International Space Station.

WHAT IF THE ISS GETS HIT BY A METEOROID?

Thwack! One of the seven windows of the space station's Cupola module was struck by a micrometeoroid in 2012. A protective shutter immediately descended to seal the window in case a leak caused a sudden station depressurization. Thanks to the window's four-layered panes, there was no breach, and the shutter was lifted. A whole window can be replaced if damaged beyond repair.

The hull of the space station is reinforced and can take hits from objects up to ⅜ inch (1 cm) across. Anything larger, however, could pierce the shell and compromise the airtight environment inside. In 2013, an impactor narrowly missed the hull and dented a solar panel. There are more chances of ISS being struck from some directions than others. It probably won't get direct hits from below, because Earth acts as a shield.

Mission controllers monitor a safety zone around the ISS that stretches to a couple of miles. Any object entering the region with odds of a collision less than 1 in 10,000 prompts an evasive change in the ISS's orbit. This has already happened more twenty times, although sometimes the offending item has been a piece of space junk instead (see pp. 108–111). Astronauts can always evacuate if there is no time to move out of the way.

v NASA astronaut Tracy Caldwell Dyson in the ISS's Cupola module; one of its windows was damaged by a micrometeoroid in 2012.

TOILETS

YOU'D THINK THE MOST COMMON QUESTION ASKED OF RETURNING ASTRONAUTS WOULD BE ON THE VIEW, BUT INEVITABLY IT'S THE ONE ABOUT GOING TO THE BATHROOM IN SPACE.

The everyday human activity of using the john comes with extra complications—and even dangers—for people in the unique environment of a spacecraft.

When you enter the telephone booth-size space toilet in the US module of the International Space Station (ISS), it doesn't look all that different from the facilities you'd find on a typical commercial aircraft. However, the price tag is much greater—it cost $30 million to build. On the wall behind the toilet is a mission patch depicting a cartoon astronaut spacewalking with a roll of toilet paper and the words "International Space Station Orbital Outhouse Team."

The challenges presented by a lack of gravity mean that astronauts have different ways of using the toilet, depending on whether their output is liquid or solid. For liquids, they unmount a long hose from the wall and turn a rotary switch ninety degrees to start the suction. Men have to be careful that they don't inadvertently "dock" with the funnel. You pee into a yellow cone at the top and it all gets drawn down into the Urine Processing Assembly (UPA) under the floor for recycling (see pp. 60–63). The dangers of stray pee were underlined in 1963 when US astronaut Gordon Cooper's urine collection system failed and the leak caused sensitive electronics to fail. He had to take manual control and fly his capsule down into the atmosphere himself during reentry.

∧ The patch that sits behind the space toilet in the American end of the International Space Station.

↗ One of the toilets on the ISS. The yellow hose hanging on the right-hand wall is for liquid waste.

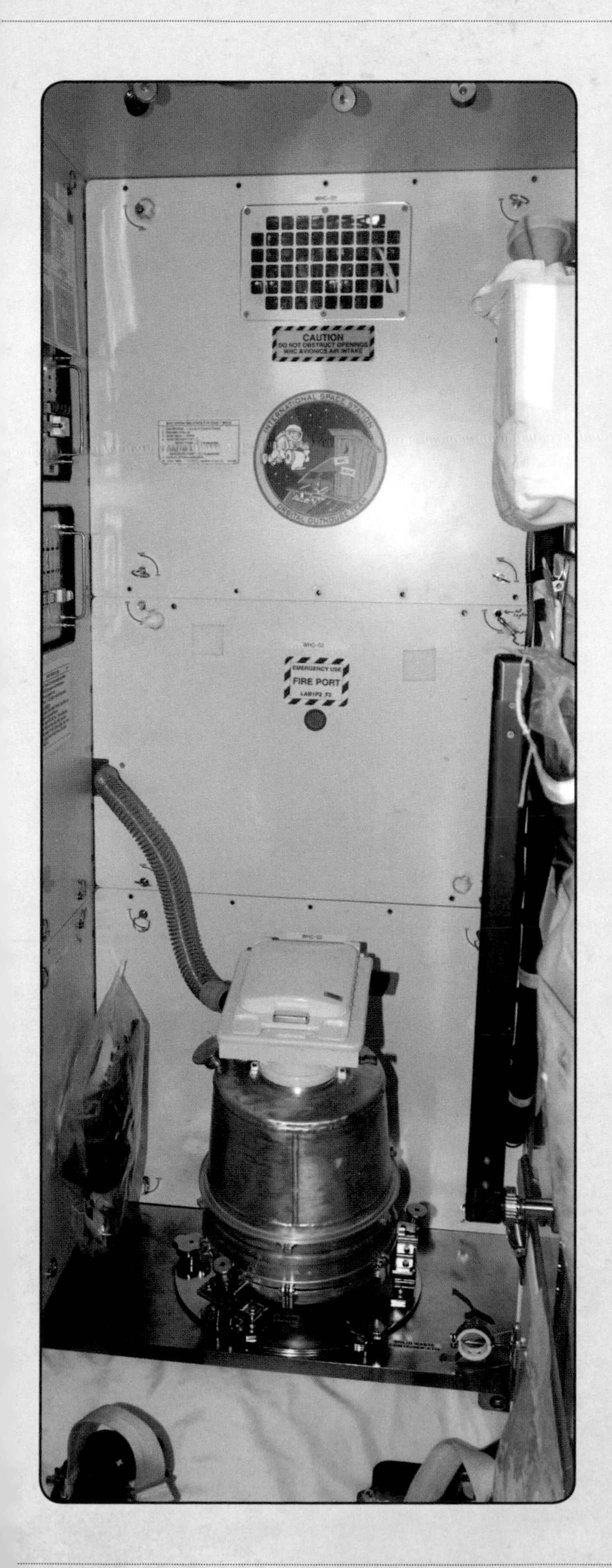

WHAT HAPPENS WHEN THE TOILET BREAKS?

Just as here on Earth, sometimes the plumbing fails, but in space you can't call someone out to fix it. A legendary story from the Space Shuttle Discovery in 1989 hammers home how bad the situation can get.

Astronaut Fred Gregory had just made a deposit, but then the toilet mechanism broke. The resulting klaxon woke his crew mates. The situation was serious—the jammed gear mechanism caused the cabin to depressurize. The toilet was out-of-bounds while calm was restored. The astronauts had to resort to the adhesive bags of the Apollo era (see box, p. 99). This led to Mission Control contemplating an emergency landing in Africa, but the fault was fixed.

In July 2009, one of the two toilets on the ISS broke. This was bad timing, given that there were a record-breaking thirteen people on board at the time. It had been installed only eight months earlier. The astronauts had to use the facilities in the Russian Zvezda service module, or the toilet in the docked Space Shuttle Endeavour. The Russian toilet also broke, in 2008, when a pump failed and astronauts had to wait for a new one to be delivered on the shuttle mission *Discovery* (STS-124).

v The $23 million improved Waste Collection System (WCS) toilet designed for use aboard Endeavour.

"Number twos" are a little more difficult. There's a solid waste container with a bag inside and a seat on top. Not that you can really sit in space, so a lot of astronauts lift the seat to directly expose the opening of the bag. That means they have a bigger hole to aim at, but it's still pretty small. On the Space Shuttle it was just 4 inches (10 cm) in diameter. There are some foot straps to hold you in place and stop you from floating away. Astronauts have come a long way from the more primitive solutions used on earlier missions, including the Apollo missions to the Moon (see box, opposite).

Once you've finished, you push the self-sealing bag down into the bottom part of the solid waste container and fit a new bag, ready for the next astronaut. It typically takes three astronauts 10 days to fill it up, and the day of the last change is written in pen on the front so they can keep track. The solid waste is then exposed to the vacuum of space—sterilization by rapid freeze-drying. Later, it is added to an uncrewed service module that detaches from the ISS and burns up in the atmosphere. So, occasionally, a shooting star is made of space poop.

v NASA astronaut Curtis Brown prepares to clean the toilet of the Space Shuttle Endeavour.

v One of the "Fecal Collection Assembly" bags used during the Apollo era.

The Apollo astronauts may have been the first to visit the Moon, but their lavatory situation was far from glamorous. To dispense with solid waste, they had to tape a plastic bag to their buttocks and do the business. A tablet inside killed bacteria and stopped gases from building up, but not everything made it into the bag.

Apollo 11 may have generated the most iconic words ever spoken in space, but the transcript of the Apollo 10 flight is far more humorous. Commander Tom Stafford: "Oh, who did it? There's a turd floating through the air." Command Module pilot John Young is apparently not guilty. "It ain't one of mine," he says. Stafford is more certain: "Mine was a little more sticky than that." Gene Cernan was even more forthright: "I sure would know if I was s***ing on the floor." Whose business it was remains unclear.

What happened to the poop of the Apollo astronauts who did make it to the lunar surface is more clear; they left it there. To this day, there are ninety-six bags of human waste on the Moon. One day, scientists could collect it to see how it's been affected by decades of exposure to space.

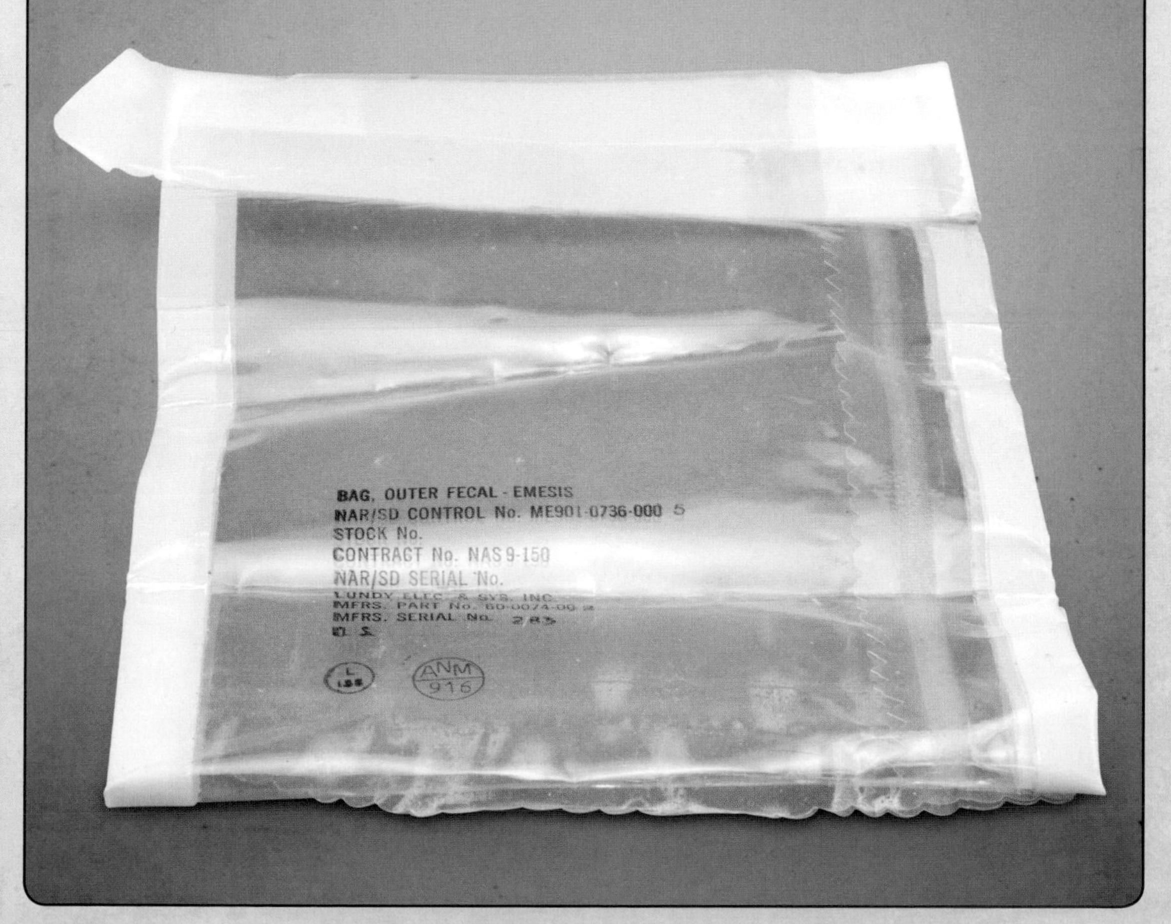

GENERATING POWER

GENERATING ELECTRICITY IN SPACE IS ALL ABOUT HARNESSING THE POWER OF THE SUN. THE INTERNATIONAL SPACE STATION (ISS) IS POWERED BY SOLAR PANELS.

The ISS has eight solar arrays—four at each end—that drink in sunlight whenever they are illuminated. Originally the station had only two, but more were added by some of the first Space Shuttle missions to visit. They were designed to fold up small inside the shuttle like an accordion, but then unfold in space until each one stretched to 115 feet (35 m) long and almost 39 feet (12 m) across—when combined, that's approaching about an acre of solar panels. Together, they contain more than two-hundred-and-fifty thousand individual solar cells and are connected by 8 miles (13 km) of electrical wiring. They are able to generate between 84 and 120 kilowatts (kW) electricity, enough to power forty homes on Earth. Their reflectivity makes the ISS easily visible from the ground (see box, right).

With the ISS speeding around the planet every hour and a half, regularly plunging in and out of darkness, about 60 percent of the electricity generated by the panels is stored in batteries, ready to be used when the Sun is obscured by our planet. During a spacewalk in 2017, some of the oldest nickel-hydrogen batteries were replaced with more efficient, modern lithium-ion batteries. A pivoting device known as a gimbal is used to angle the ISS constantly toward the light. The current generated is direct current (DC) instead of the alternating current (AC) you find in the typical domestic electricity supply.

∧ NASA astronaut Scott Parazynski repairing a damaged solar array on the International Space Station during a spacewalk.

> NASA astronaut Scott Kelly took this image of dancing aurorae behind the ISS's solar panels in 2015.

SEEING THE INTERNATIONAL SPACE STATION FROM EARTH

Thanks to its wide array of shiny solar panels, the ISS is incredibly reflective and can easily be seen in the night sky when passing overhead. Its trajectory carries it over 90 percent of the planet's population, and there are several places online where you can easily find out when it's due over your location.

Astronomers measure the brightness of objects in the sky in terms of their apparent magnitude. The lower the number, the more dazzling the object. Sirius —the brightest star in the night sky—has an apparent magnitude of -1.47. The planets Jupiter (-2.94) and Venus (-4.89) are even brighter. Yet, at its closest point to Earth, and when completely lit by the Sun, the ISS shines with an apparent magnitude of -5.9.

You'll see it as a rapidly moving, constant white light, a lot like an airplane but much higher and without the flashing lights. It'll rise from the horizon before disappearing into Earth's shadow toward the top of the sky. However, don't expect it to be in your sky for long. Remember it takes just ninety-two minutes to orbit Earth, so it will be above you for only a few minutes.

Sometimes, however, the lights go out. The Space Shuttle Atlantis took off from Florida in 2007 to deliver a new array of solar panels to the ISS. Astronauts undertook a series of spacewalks during the mission to attach the panels to the starboard side of the station. However, soon after the panels received their first beams of sunlight, a surge of electrical current fried one of the Russian electrical boxes on board. Three of the Russian control and command computers immediately shut down.

To work out the source of the problem, Mission Control ordered the crew to turn off all computers, electrical systems, and science experiments. The ISS was plunged into darkness, with the astronauts resorting to flashlights to read manuals. Not only was it dark and eerie, but carbon dioxide (CO_2) levels were rising and the primary oxygen generator was offline.

The problem was eventually fixed by hot-wiring the computers. The astronauts used jumper cables to bypass the soft switches designed to protect the system from power surges. Further investigation found that the installation of the solar panels was coincidental and the boxes were probably fried by an encounter with the electrically charged plasma that the ISS flies through as it orbits Earth.

< NASA's New Horizons mission to Pluto is one in a long line of space probes to be powered by Radioisotope Thermoelectric Generators (RTGs).

NUCLEAR POWER

When it comes to robotic space exploration, often the energy producer of choice is a radioisotope thermoelectric generator (RTG). These have been used for decades on many of the flagship missions to explore the Solar System, including the Curiosity Mars rover, the Cassini mission to Saturn, and the New Horizons probe sent to flyby Pluto. Generating energy through the gradual radioactive decay of elements, such as plutonium-238, they are small and lightweight and complement a probe's own solar panels, particularly when journeying far from the Sun and experiencing ever-dwindling light levels.

You'll find them aboard the Voyager probes launched to explore the outer Solar System in 1977, and they are still working today. Mission scientists think they'll hold out until about 2025, a staggering shelf life. Unfortunately, the same technology cannot be used on human space missions, because the shielding required to protect the astronauts from the radiation created would be unfeasibly heavy. NASA is looking at a nuclear fission generator instead, testing an early prototype at the beginning of 2018. It could be used on the first human missions to Mars later this century (see pp. 148–171), generating 40 to 50 kilowatts of power.

v NASA's Kilopower Reactor Using Stirling Technology (KRUSTY) experiment to test out nuclear fission engines.

WE SO OFTEN HOLD ASTRONAUTS UP AS HEROES THAT IT IS SOMETIMES EASY TO FORGET THEY ARE JUST LIKE EVERYBODY ELSE AND SUBJECT TO THE SAME STRESSES AND STRAINS ON THEIR MENTAL HEALTH.

So space agencies are careful to look after their crews' psychological as well as physical needs during missions.

Six months is a long time to be away from home, cooped up in a football field-size metal can with the same people, which is why the field of aerospace psychiatry has been growing. This discipline can trace its origins back to the Soviet Salyut space stations in the 1970s. These were some of the first missions to spend many months in orbit, but a few had to be curtailed, because the crew were experiencing difficulties.

Today, astronauts aboard the International Space Station (ISS) are looked after by a dedicated team of psychologists. They begin working with the astronauts up to two years before launch, and, importantly, the space traveler's spouse and wider family are part of the program. Six months away can be an adjustment for them, too. That's why it's crucial that the astronauts can readily stay in touch with them from space (see box, p. 107). Once on the ISS, each crew member has a private video conference every two weeks with the psychology team to address any

DEAR DIARY

Between 2003 and 2016, 20 astronauts were asked to record their daily activities in the form of diary entries in an attempt to understand their behaviors and thoughts. They generated half-a-million words—enough to fill two 1,000-page books. The astronauts spent an average of 174 days in space, meaning that their collective journals covered the equivalent of 10 years in orbit.

Researchers analyzed their writings for factors that contributed to both successes and difficulties. Statements were put into positive, negative, or neutral categories, and also depending on the task they were writing about. Interestingly, astronauts found the act of noting down their thoughts useful for gaining perspective on events. They show that conditions on board the ISS were above a level considered tolerable, but there were notable instances of frustration and annoyance. On the whole, life on the station was easier than the astronauts had expected prior to launch. However, it was also found that procedures on board were not working as optimally as they could be, so the results are being fed back into new schedules and practices designed to make life better and more efficient for the crew.

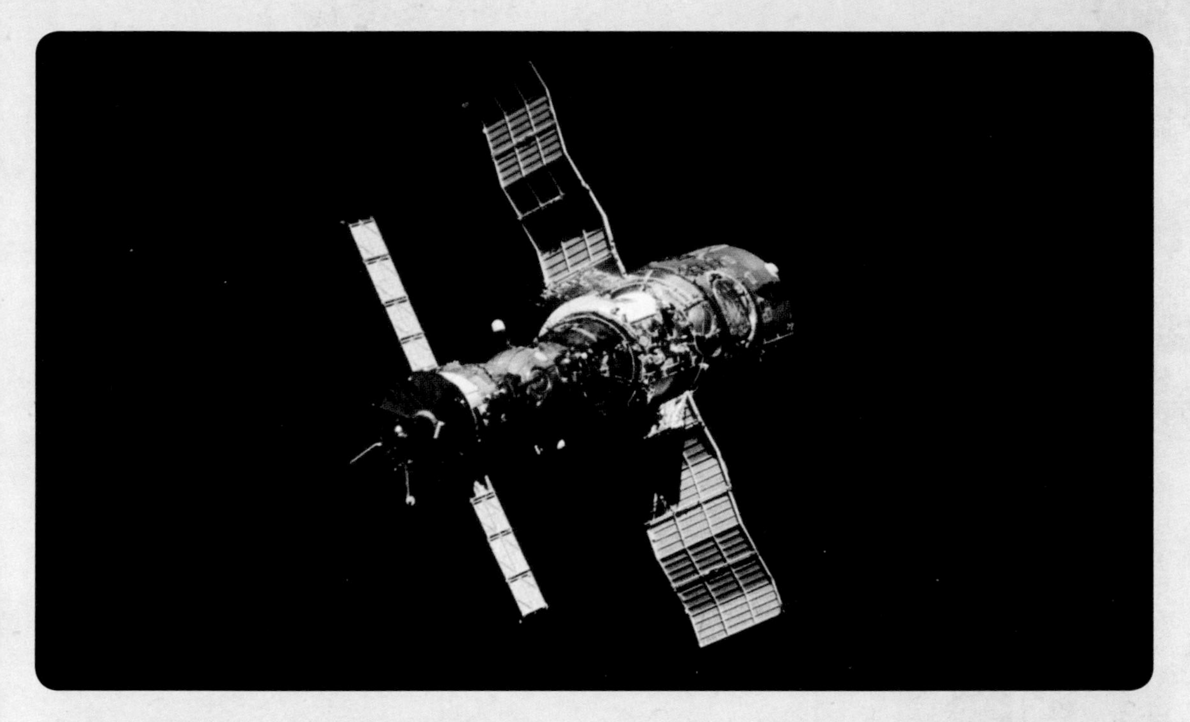

∧ The Soyuz T-4 spacecraft docked with the Soviet space station Salyut 6. The photo was taken from the Soyuz 40 spacecraft.

v NASA astronaut Peggy Whitson inside the cargo bag she hid in as a prank on her fellow crew members.

concerns. They have the team's direct telephone number and email address in case they need help at any time. After returning to Earth, they undergo routine psychological assessments three days, fourteen days, and then thirty to forty-five days after touching down. Diaries are also kept to help monitor the astronauts' mental state (see box, p. 104).

A key factor in keeping astronauts sane is free time. Space agencies want to get as much bang for their buck as possible in terms of astronaut output, but they recognize the benefits of some time for rest and relaxation. A healthy amount of downtime is built into their schedule. Astronauts can watch movies, read, play cards, or indulge in their absolute favorite pastime: staring out of the window at Earth. They often share their photographs on social media. Astronauts even play pranks on each other. In 2017, NASA astronaut Peggy Whitson packed herself into a cargo bag and jumped out to surprise the three Russian cosmonauts on board. A year earlier, Scott Kelly dressed up in a gorilla suit and chased his colleagues around the ISS.

Crew care packages are also a real morale booster. A resupply ship will typically dock with the ISS once or twice during a six-month stay. The packages measure just 9 by 16½ inches (23 x 42 cm), and can weigh only 11 pounds (5 kg), but they often contain items that are precious to the astronauts. Included could be a drawing from your child or an all-important resupply of chocolate rations. Understanding the benefits of these perks to mental health will be invaluable when planning trips deeper into space (see p. 148).

∧ NASA astronaut Scott Kelly inside a gorilla suit he wore while chasing other astronauts around the ISS.

∨ NASA astronaut Greg Chamitoff with a chessboard on the ISS. He was playing against the public on Earth.

STAYING IN TOUCH WITH FRIENDS AND FAMILY

One of the obvious impacts of long stays in space is that you're far removed from your nearest and dearest. Astronauts often cite it as one of the hardest parts of the job.

To minimize the effect on the astronauts' psyche, there's an Internet phone installed in the ISS. Astronauts speak through a headset. They don't need to get permission to dial out and can call anyone they want; however, the long distances involved means there's a one-second lag in the conversation. In case you're wondering, the number that shows up on your phone is 000000.

Astronauts also have regular video calls with their family. They can be important for keeping up with birthdays and anniversaries. Occasionally, space travelers miss the birth of their children, as happened with NASA astronaut Michael Fincke in 2004. He phoned his wife in the delivery room, but didn't meet his new daughter until she was four months old. Unfortunately, astronauts can also miss the death of loved ones. NASA astronaut Daniel Tani's mother died in a car accident during his mission in 2007. These events make keeping the connection to Earth a crucial part of space travel.

v Paolo Nespoli (left), Cady Coleman (middle), and Scott Kelly (right) unwrap a care package delivered to the ISS.

SPACE JUNK

SINCE SPUTNIK 1 WAS SENT INTO SPACE IN 1957, WE'VE CONDUCTED MORE THAN FIVE THOUSAND LAUNCHES, DISPATCHING MORE THAN SEVEN THOUSAND SATELLITES INTO ORBIT.

This activity has led to unprecedented levels of space pollution. Our gateway to space is becoming crowded, and many space scientists are worried that the consequences of a collision could be catastrophic (see box, p. 111).

Fragmented items left behind in orbit—space junk—include discarded covers, rocket boosters, paint flakes, and frozen coolant. There is thought to be 8,270 tons (7,500 tonnes) of human-made material in orbit, comprised of

THE MIR SPACE STATION

Not all space junk is small. Nor does it always exclusively threaten spacecraft and the astronauts inside. Sometimes satellites lose energy, their orbit decays, and they fall to the ground. This can be a deliberate act by space scientists, or an unintended consequence of a natural phenomena, such as a solar storm that causes our planet's atmosphere to expand and suddenly increase friction with a satellite.

The biggest object to fall from orbit was the Russian station Mir—the first continuously inhabited long-term research outpost in space. Controllers deliberately deorbited the station in March 2001. It reentered the atmosphere and broke up close to Fiji in the Pacific Ocean. While most pieces were incinerated, some large parts did reach the surface and were sprayed across a corridor measuring 930 by 62 miles (1,500 x 100 km). Both New Zealand and Japan issued warnings to sailors, pilots, and residents ahead of the exercise. The chances of anyone being struck were small.

One day, the ISS will also have to come down, either due to age or because of an emergency. If the ISS ever has to be completely abandoned following an enforced evacuation, there's a fourteen-day window for Mission Control to decide whether to bring it down over the open ocean.

both functional and nonfunctional hardware. The US Space Surveillance Network regularly tracks more than twenty-three thousand orbiting items. Estimates suggest that there are more than twenty-nine thousand items of space junk more than 4 inches (10 cm) across, seventy-five thousand between ½ inch and 4 inches (1–10 cm) and nearly two-hundred million between 1⁄25 inch and ⅜ inch (1 mm–1 cm).

Orbital debris travels eight times faster than a bullet, so a small collision can do sizable damage to space stations and spacewalking astronauts. In 2016, a tiny fleck of paint chipped a window of the International Space Station (ISS). Fortunately, the portals are made from several layers for just such an event.

The problem is becoming so acute that mission controllers added a Space Debris Sensor (SDS) to the

∧ A model of the space junk around Earth, based on real data but not shown to scale.

< An artist's impression of orbital debris after a collision between satellites.

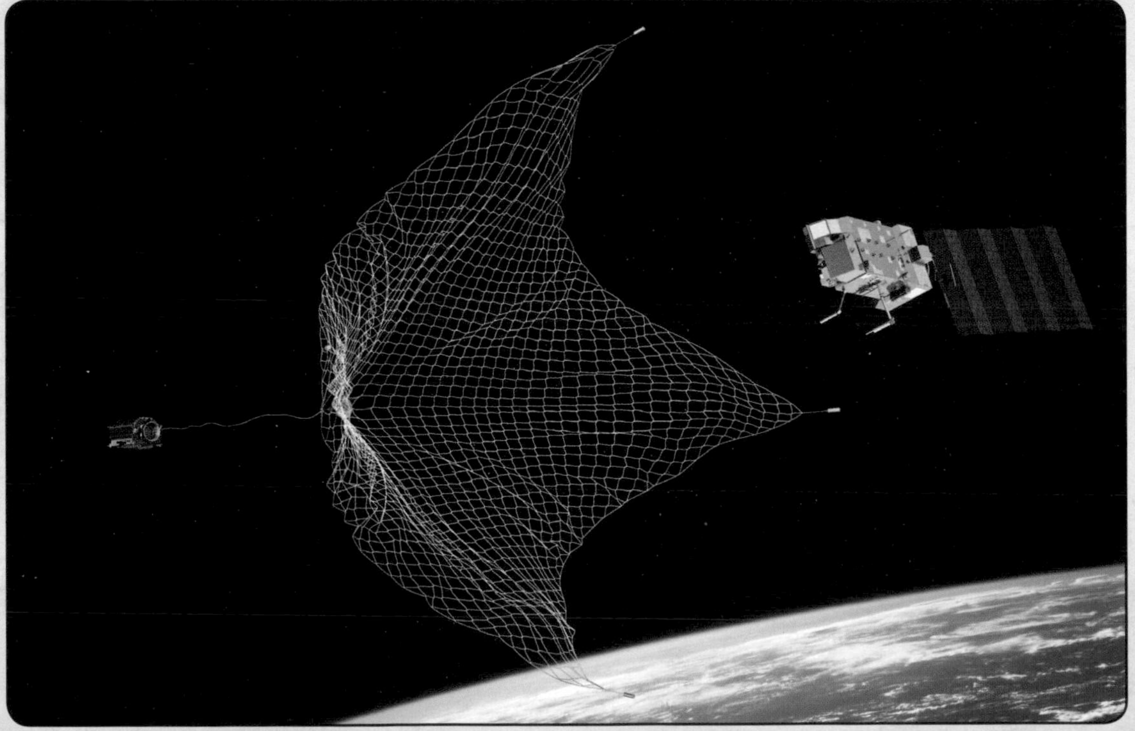

ISS in September 2017. Radar surveys from Earth simply can't pick up a enough number of the smaller objects. They are most worried about steel fragments just 1–1.5 millimeters across, because these have the capability to pierce the hull of the station, leading to a dangerous depressurization. The SDS has two layers of Kapton film separated by 6 inches (15 cm). The time delay between each layer being broken can be used to work out the speed of the impactor. Changes in the films' resistance are used to estimate the impactor's size.

New rules are supposed to ensure that a satellite safely deorbits itself after twenty-five years, however, that's proving difficult to enforce. It doesn't help when countries, such as China, deliberately destroy one of their satellites with a missile, as happened in a 2007 show of strength. The impact created an estimated two-and-a-half thousand pieces of additional debris that could threaten other space technology for decades to come. So scientists are trialing technology that could help with a grand cleanup. In April 2018, the RemoveDebris spacecraft was launched to the International Space Station. Once there, the resident astronauts will deploy it into orbit. Built in the UK, the washing-machine-size experiment will release its own little pieces of space junk before attempting to recapture them. Methods that it will test include using a net to snare the junk and a harpoon to impale it.

Mounted on the exterior of the ISS, the Space Debris Sensor (SDS) collects information on small orbital debris.

< A net could be used to snare space junk and drag it into the atmosphere to burn up safely.

v This hole was made in the SolarMax telescope by a piece of orbital debris.

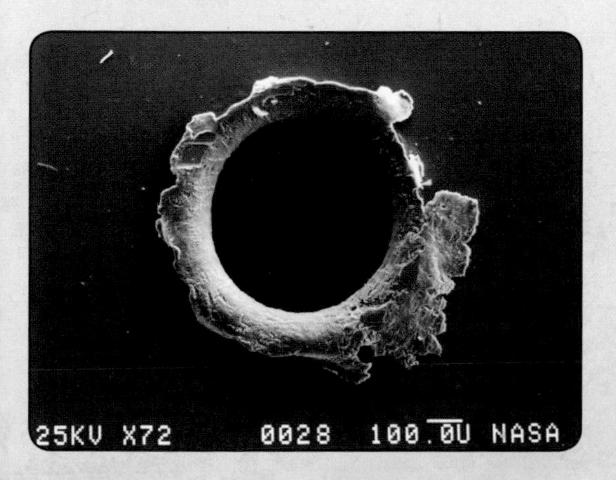

v The ESA's *Envisat* environmental satellite is a big concern for space junk analysts.

How bad could the space junk problem get? The answer is catastrophic. In 1978, NASA scientist Donald Kessler published a study into what has now become known as Kessler syndrome.

As the density of space junk rises, so does the risk of collisions. One crash would trigger another and then another. Like a ping-pong ball dropped into a box of mousetraps, a cascade of events would follow that would clear out large swathes of low Earth orbit (LEO). That could cripple our activities in space for many decades. We've already had warnings. In 2009, a communications satellite crashed into a defunct Russian spacecraft. In 2017, a computer simulation run by the University of Southampton in England found that future launches could raise the risk of catastrophic collisions by 50 percent over the next two hundred years.

Kessler and others have singled out the danger posed, in particular, by one large satellite: the European Space Agency's Envisat. Launched as an Earth observation station, it almost collided with a Chinese rocket in 2010. There's a 15 to 30 percent chance of another collision over the one-hundred-and-fifty years it will probably take for the satellite's orbit to decay naturally, bringing it into Earth's atmosphere.

SPACE LAW

IT IS A TRUTH, UNIVERSALLY ACKNOWLEDGED, THAT A SINGLE COUNTRY IN POSSESSION OF A ROCKET MUST BE IN WANT OF A SPACE LAWYER.

Humanity's first steps into space have produced a host of possible issues to do with utilizing space for the benefit of everyone.

The 1957 launch of Sputnik 1 by the Soviet Union had many people running for the legal textbooks and lawyers' offices. It was such a revolution in human achievement that no one had really thought about the legal implications. Traditionally, a country owned the airspace above its sovereign borders. So, does another nation have to ask for permission every time it wants to fly a satellite over your land? What if it crashes? In 1979, a piece of the US Skylab space station landed in Western Australia. The local government issued NASA a $400 fine for littering.

To address this, and the many other legal and diplomatic challenges presented by the newly minted space age, the United Nations General Assembly convened the Committee on the Peaceful Uses of Outer Space (COPUOS) in 1959. Initially, it had twenty-four member states, but today that number has

WHAT HAPPENS IF A CRIME IS COMMITTED IN SPACE?

With astronauts living together in close quarters, it's inevitable that tensions in space will run high at times. But what would happen if one astronaut assaulted another? Or worse, murdered a crew mate?

If the spacecraft belongs to one nation, then the answer is simple. The assailant would be arrested and tried according to that country's laws, just as the laws of the operating airline's registered country apply on commercial flights, or the laws for a ship's country of registration apply to a vessel in international waters. On the International Space Station (ISS), the situation is clearly explained by the legal agreement signed in 1998 by the cooperating nations. The crime falls under the jurisdiction of the perpetrator's country. If they fail to press charges, the victim's country can exercise jurisdiction.

What happens if the crime is committed on the Moon or Mars? The 1967 Outer Space Treaty specifically forbids one country from claiming sovereignty over these bodies. So whose laws apply? Here, the principle of "extraterritorial jurisdiction" rules. An astronaut is bound by the laws of their home country, even when beyond its borders.

While current astronauts are highly trained professionals, a future in which we see more space tourists (see pp. 124–135) could lead to this legal framework being tested.

< The first meeting of the Ad Hoc United Nations Committee on the Peaceful Uses of Outer Space (1959).

v The first international treaty on outer space opened for signature in 1967.

v A map of the world showing which nations have agreed to the 1967 Outer Space Treaty.

swelled to more than eighty, with additional observer status for some international organizations. There's a subcommittee of COPUOS called the United Nations Office for Outer Space Affairs (UNOOSA). In the past this subcommittee has considered introducing official protocols for what to do if we ever make contact with alien life.

The early work of COPUOS laid the framework for more modern international space law. Existing legislation was drawn together in the late 1960s into what is known as the 1967 UN Outer Space Treaty. Its full name is the *Treaty on Principles Governing the Activities of States in the Exploration and Use of Outer Space, including the Moon and Other Celestial Bodies*. More than a hundred countries are now party to it.

NATIONS SIGNED UP TO THE 1967 UN OUTER SPACE TREATY

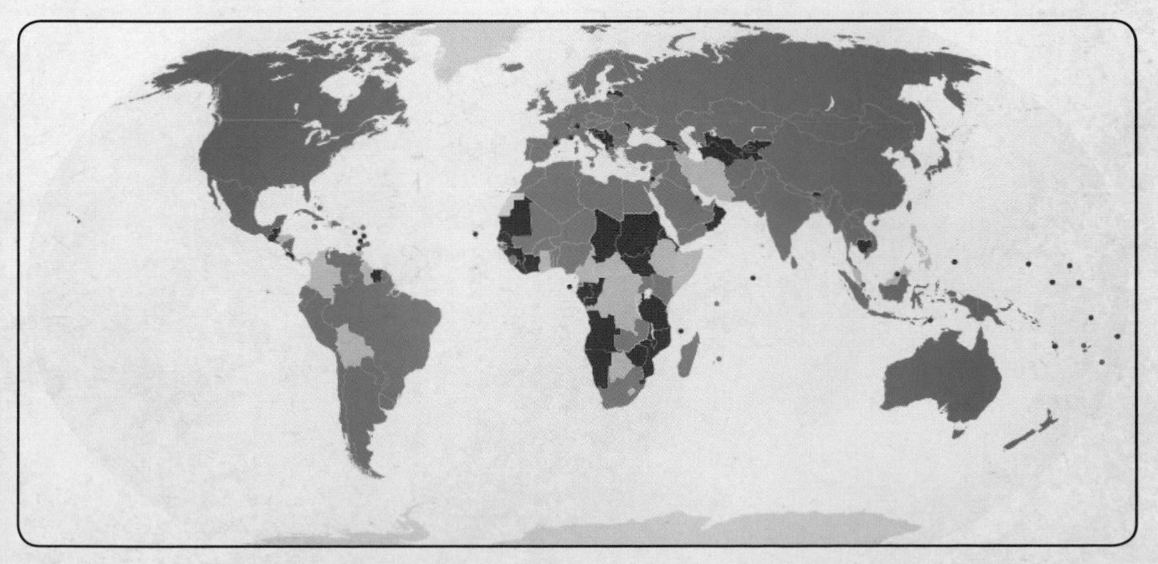

PARTIES　SIGNATORIES　NONPARTIES

It enshrines the following principles into law:

- The exploration and use of outer space will be carried out for the benefit and in the interests of all countries and will be the province of all mankind
- Outer space will be free for exploration and use by all states
- Outer space is not subject to national appropriation by claim of sovereignty, by means of use or occupation, or by any other means
- States will not place nuclear weapons or other weapons of mass destruction in orbit or on celestial bodies, or station them in outer space in any other manner
- The Moon and other celestial bodies will be used exclusively for peaceful purposes
- Astronauts will be regarded as the envoys of mankind
- States will be responsible for national space activities whether carried out by governmental or nongovernmental entities
- States will be liable for damage caused by their space objects and states will avoid harmful contamination of space and celestial bodies.

PROTECTING OUR SPACE HERITAGE

> Buzz Aldrin during the *Apollo 11* Extra-vehicular Activity (EVA) on the Moon in 1969.

Since 1978, the United Nations Educational, Scientific, and Cultural Organization (UNESCO) has been adding to a list of internationally recognized World Heritage Sites—places that mark a special place in our history and therefore need protecting. More than 190 countries have signed up to the convention. But what about the six Apollo landing sites on the Moon that still host flags and footprints?

Current space law makes this more than a gray area. According to the 1967 UN Outer Space Treaty, the US government would find itself in hot diplomatic water if it unilaterally declared these sites off-limits. The flags remain US property, but the footprints form part of the lunar surface, and that cannot be claimed by a country. A bill—the Apollo Lunar Landing Legacy Act—was introduced in the US Congress in 2013 with the aim of making the Apollo landing sites off-world national parks. More recently, an organization called For All Moonkind has been drawing up a protection plan that they hope to present to the UN Committee on the Peaceful Uses of Outer Space. Such an initiative is crucial if we're to protect our space heritage in the coming age of commercial space travel.

REENTRY

A HOVER OF HELICOPTERS SCOURS THE DRY STEPPES OF KAZAKHSTAN WHILE SPACE SCIENTISTS ARE BUSY CRUNCHING NUMBERS. THESE EFFORTS ARE PART OF YOUR RETURN FROM ORBIT.

Your landing site has been precisely calculated and the helicopters are there to reconnoiter the area for any obstructions.

With a week still to go, you begin a remote training session with mission controllers to refamiliarize yourself with landing procedures and the all-important emergency operations. As the day of descent arrives, you don't have time to daydream of home, because you've got a ton of preflight checks to conduct. By the time you close the hatch linking the Soyuz to the International Space Station (ISS), you're just three-and-a-half hours from terra firma.

The hooks holding your craft to the space station gradually disengage over several minutes as the *Soyuz* slowly backs away from the ISS at a speed of just 5 inches (12 cm) per second. Firing the capsule's rockets could damage or contaminate the ISS. At a safe distance, you perform a series of engine burns, first to increase your separation from the ISS, then to decrease your speed to enter the deorbit trajectory. The timing of this burn is crucial. If you come in at too shallow an angle, you'll simply skip off the atmosphere. Come in too steeply, and the temperature will climb so high that the Soyuz's heat shield could be compromised. For these reasons, the deorbit burn lasts a precise four minutes and forty-five seconds.

You're now under an hour from landing, and already gravity is starting to grip you hard for the first time in

^ Russian Search and Rescue and other recovery teams begin to gather around the Soyuz TMA-20 spacecraft after its 2011 landing.

< The Soyuz TMA-21 spacecraft landing in 2011 in a remote area outside of the town of Zhezkazgan, Kazakhstan.

^ NASA astronaut Chris Cassidy is carried to the medical tent shortly after landing in 2013.

ADJUSTING TO LIFE BACK ON EARTH

When you've been in space for six months, gravity is a drag—literally and metaphorically. Your body has adapted to weightlessness and now finds itself perpetually being pulled downward. Adjustments are often relatively simple. When Canadian astronaut Chris Hadfield came back down in 2013, he quickly learned to speak again—after half a year of a weightless tongue had shifted the way his mouth and lips formed words.

All returning astronauts undergo a barrage of medical examinations, both to check that they're okay and to learn more about the long-term effects of space travel. Vision problems are common, because gravity changes the pressure inside your eyeballs. After Scott Kelly's year in space (see pp. 68–71) he initially experienced a burning sensation whenever he tried to sit or lie down; his skin hadn't had pressure on it for so long that it had become supersensitive. You're banned from driving for at least three weeks as you readjust to life back on the surface of a planet.

Having a proper shower must be a real pleasure, except that most astronauts take one sitting in the tub at first, because of the risk of slipping due to balance issues. NASA has also introduced a policy of seating astronauts at post-flight press conferences after Heide Stefanyshyn-Piper fainted twice while trying to stand.

months. Temperatures outside climb toward a scorching 2,000 degrees Celsius as you tear through the upper atmosphere. With thirty minutes to go, and at an altitude of 87 miles (140 km), explosive bolts fire to separate your descent module from the other two sections of the *Soyuz*, which burn up in the atmosphere. When you're just 6½ miles (10.5 km) above the ground, a series of parachutes are released to slow you down further—an automatic step with no manual override. The 10,750-square-foot (1,000-sq-m) main parachute is deployed, and excess fuel is vented to reduce the risk of exploding on landing. At 28 inches (70 cm) from touchdown, six retro rockets fire to reduce your speed to 3 miles (5 km) per hour. This is a so-called "soft landing," but any astronaut would balk at that description. Shock absorbers cushion some of the impact, but astronauts adopt a brace position and there is strictly no talking—you wouldn't want to jam your tongue between your teeth on impact. Finally, helicopters deliver the rescue crew to the landing site to open the hatch and carry your fragile, gravity-crippled body out to safety.

v The different stages of a Soyuz landing, from module separation to touchdown.

THE SOYUZ 11 AND SPACE SHUTTLE COLUMBIA DISASTERS

v Debris from the *Columbia* disaster is organized inside an aircraft hanger by investigators.

Reentry remains one of the most dangerous parts of any space mission. Unfortunately, it has led to fatal accidents.

Georgy Dobrovolsky, Vladislav Volkov, and Viktor Patsayev—the astronauts of the 1971 *Soyuz 11* mission—are the only people ever to die in space, when returning from the Salyut 1 space station. To add to their misfortune, they had replaced the first crew at only the eleventh hour, because an X-ray had hinted that one of the original astronauts might have tuberculosis. The reentry itself went off without a hitch, but when the hatch was opened upon landing the crew were found dead. Further investigation revealed that they'd suffocated, because their descent capsule had depressurized on separating from the orbital module.

In 2003, the Space Shuttle *Columbia* disintegrated during reentry, killing all seven astronauts on board. They were beneath the Kármán line (see p. 8) at the time. A piece of insulation had broken off and hit the shuttle's left wing during launch. The damage comprised the wing's ability to withstand the intense heat of reentry, and the wing broke apart. Following the crash, a huge ground search was undertaken to collect debris, including some of the human remains, strewn over two thousand sites across Texas, Louisiana, and Arkansas.

A TIMELINE OF ROBOTIC SPACE EXPLORATION

1940s

October 3, 1942—a German V2 rocket launched during the Second World War becomes the first human-made object to cross the Kármán line.

1950s

October 4, 1957—the Soviet Union launches the Sputnik 1 satellite into orbit—kick-starting the space race, a battle to dominate this new frontier.

1960s

Spacecraft are sent to explore the Moon and our neighboring planets Venus and Mars, taking the first up-close photographs of other worlds.

1970s

Robotic space probes return rocks from the Moon, the Viking landers test Martian soil for life and *Pioneer 10* moves out beyond the Asteroid Belt for the first time.

1980s

The Soviet Venera 13 mission records sounds on Venus and NASA's Voyager 2 reaches Uranus and Neptune, having already explored Saturn earlier in the decade.

1990s

NASA's Galileo probe performs the first asteroid flyby before reaching Jupiter. Soujorner becomes the first rover on Mars.

2000s

The Cassini-Huygens mission reaches Saturn, landing a probe on Titan. NASA's Stardust mission returns a comet sample and the Spirit and Opportunity rovers explore Mars.

2010s

Japan's Hayabusa probe returns a sample from an asteroid, MESSENGER orbits Mercury and Voyager 1 becomes first human-made object to enter interstellar space.

ISS CONFIGURATION

AS OF JUNE 2017

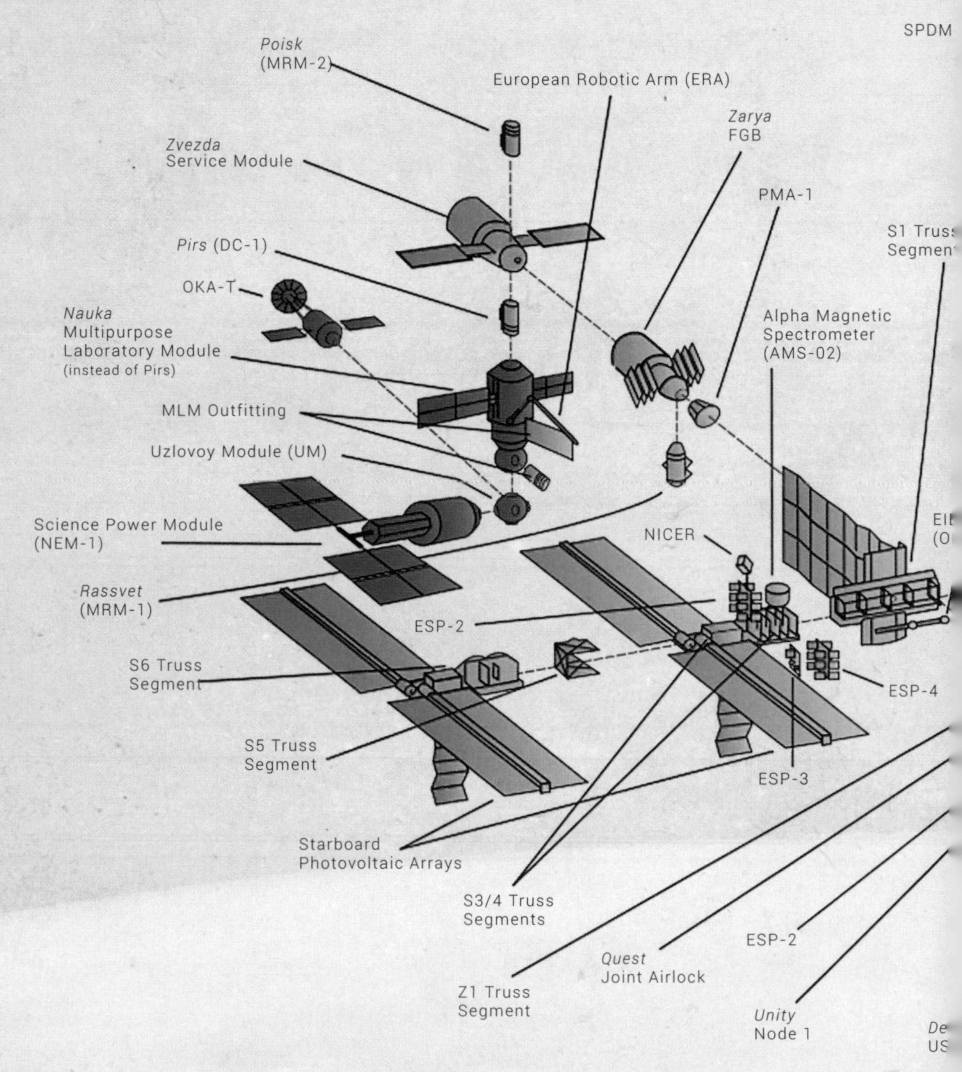

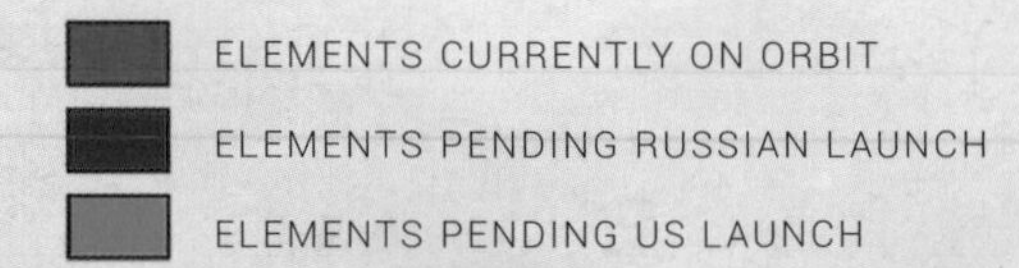

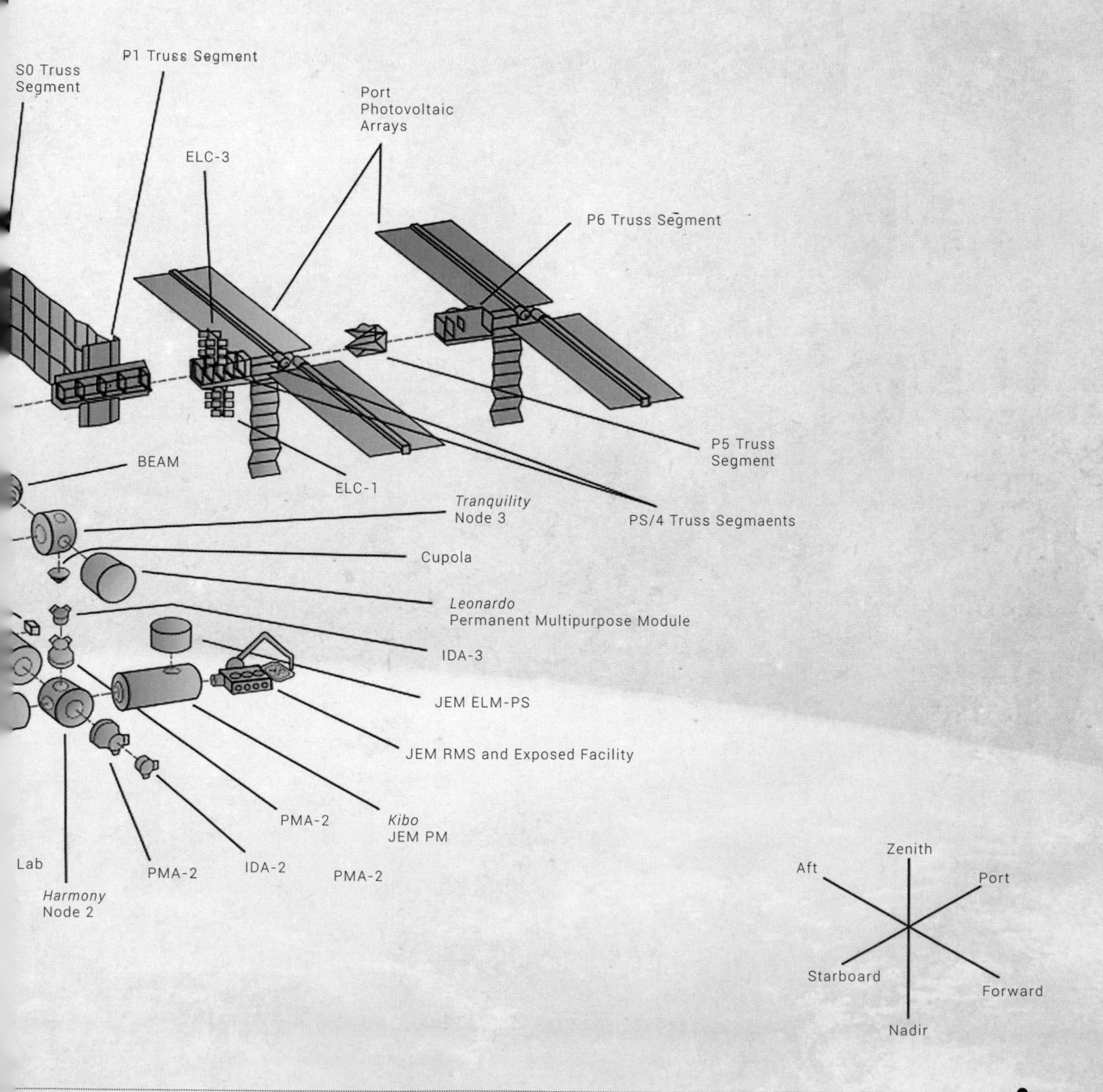

rm2
Mobile Base System
Mobile Transporter
P1 Truss Segment
S0 Truss Segment
Port Photovoltaic Arrays
ELC-3
P6 Truss Segment
P5 Truss Segment
BEAM
ELC-1
Tranquility Node 3
PS/4 Truss Segmaents
Cupola
Leonardo Permanent Multipurpose Module
IDA-3
JEM ELM-PS
JEM RMS and Exposed Facility
PMA-2
Kibo JEM PM
Lab
PMA-2
IDA-2
PMA-2
Harmony Node 2
Zenith
Aft
Port
Starboard
Forward
Nadir

HUMAN SPACE EXPLORATION

1940s

February 20, 1947 – Fruit flies become the first animals in space. Soon monkeys are going too in order to check whether it is safe for humans to follow.

1950s

A stray dog called Laïka becomes the first living creature to orbit the Earth aboard Sputnik 2 in 1957.

1960s

This landmark decade sees Yuri Gagarin becom the first human in space on April 12, 1961. Alexe Leonov later makes the first spacewalk and by 1969 astronauts are walking on the Moon.

1970s

Humans leave the Moon for the last time in 1972 and focus on life aboard orbiting space stations such as Salyut 1 and Skylab.

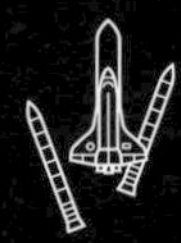

1980s

The Space Shuttle enters operation, but the program is set back by the Challenger disaster in 1986. Bruce McCandless makes the first untethered spacewalk.

1990s

International cooperation in space grows after the fall of the Soviet Union. In 1995 NASA's Space Shuttle Discovery docks with the Russian Mir space station.

2000s

Mir crashes into the Pacific as Dennis Tito becomes the first paying visitor to the newly constructed International Space Station. The Chinese launch their first taikonauts.

2010s

The Space Shuttle is retired. Lettuce is grown and eaten on the ISS and SpaceX successfully test their Falcon Heavy rocket.

PART 3

THE FUTURE

SPACE TOURISM

TOURIST FLIGHTS

AS THE SUN RISES OVER NEW MEXICO, YOU'RE FILLED WITH NERVOUS EXCITEMENT. YOU'RE SITTING ON THE RUNWAY AT SPACEPORT AMERICA—THE WORLD'S FIRST PURPOSE-BUILT COMMERCIAL GATEWAY TO THE HEAVENS.

Springing up from the surrounding desert, it covers 27 square miles (70 sq km). After three days of preflight preparation you find yourself at the end of the 12,000-foot (3.6-km) runway, ready to become an astronaut.

You're safely stowed away inside SpaceShipTwo, a vessel hanging underneath the huge wings of the mothership WhiteKnightTwo. Then, soaring into the air like a conventional commercial airliner, you find yourself carried to 50,000 feet (15,200 m), where SpaceShipTwo detaches to make the rest of the journey solo. Its powerful rockets fire and, before you know it, you're roaring through the upper atmosphere at three-and-a-half times the speed of sound. Just one minute later and you're in space. The pilot kills the power and the spaceship is silent.

The first clue you've left Earth behind will come as the familiar blue sky turns inky black out of one of the 12 huge windows. The second clue swiftly follows when you notice your seat belt struggling to keep you in your seat. A quick unbuckle and for the first time in your life you're weightless, free to float around the cabin and drink in the splendor of a panorama beyond anything you could ever have imagined. Cities are spread out before you as the planet curves away into the distance.

A few minutes later, and your stay in space is over. You return to your custom-built seat to tackle the most dangerous part of the trip: reentry. As the atmosphere

THE ANSARI X PRIZE

Sometimes a spark of inspiration is needed for us to break free from the ways things have always been. In 1919, a prize of $25,000 was offered to anyone who could successfully fly across the Atlantic Ocean. After Charles Lindbergh answered the call in 1927, commercial transatlantic aviation eventually became run-of-the-mill.

Equally, the cost needs to be reduced significantly for space to become the playground of the ordinary person. Enter the Ansari X Prize. Initially offered in 1996, $10 million was up for grabs to anyone who could "build and launch a spacecraft capable of carrying three people to 100 kilometers [60 miles] above Earth's surface, twice within two weeks." That level of reusability is key to driving down the cost by doing away with the need for new hardware each time.

The prize was won in October 2004 by SpaceShipOne, a venture backed by Microsoft billionaire Paul Allen. The technology has continued to be refined and tested in the years since. The space plane's successor—SpaceShipTwo—will carry paying tourists into space as part of entrepreneur Richard Branson's Virgin Galactic enterprise, democratizing space and giving us all the chance to experience weightlessness and see Earth curving away below us.

v A view of what it might be like inside the Virgin Galactic spacecraft.

Virgin
N348MS
V1RG1N GALACTIC
N202VG
Virgin
Virgin
N348MS
V1RG1N GALACTIC

Virgin
aabar
Virgin
N202VG

thickens, friction heats the exterior of the craft and the ride gets bumpy. Eventually, you slow enough that SpaceShipTwo becomes a glider—much like NASA's retired Space Shuttles. You coast back to Spaceport America to toast your new status as a fully fledged space traveler and receive your astronaut wings.

This scenario will soon be a reality with the first paid Virgin Galactic flights imminent. The cost of a ticket is currently set at $200,000—enough for wealthy celebrities, such as Tom Hanks, Angelina Jolie, and Ashton Kutcher, to have bought theirs already. However, the price will come down. The first commercial transatlantic jet flight took place in 1958 and cost almost $500 (aobut $4,300 in today's money). Now you can fly across the same ocean for a little as $100. If commercial space travel comes down in price at a similar pace, it will become a leisure option for everyone. The schoolchildren of today will be the space conquerors of tomorrow and will teil their grandchildren of a bygone age when space was only the preserve of a lucky few.

< SpaceShipTwo attached to the WhiteKnightTwo (top) and in space once separated (bottom).

Space tourists are not some far-flung pipe dream—they already exist and have done for years. In April 2001, American multimillionaire Dennis Tito paid $20 million to spend just over a week on board the International Space Station and become the first space tourist. He orbited Earth 128 times. A year later, South African computer entrepreneur Mark Shuttleworth followed Tito by spending nearly ten days in orbit. In September 2009, Canadian Cirque du Soleil billionaire Guy Laliberté spent twelve days in space at a cost of $35 million.

Other famous names have come close, but ultimately missed out. Lance Bass from the 1990s boy band NSYNC had hoped to go as part of a television documentary, but producers failed to raise the $20 million cost. Singer Sarah Brightman was set to visit the International Space Station and had been training in Moscow. She'd even written a special song with ex-husband Andrew Lloyd-Webber for the occasion. However, she announced in 2015 that her plans had changed due to "personal family reasons."

Soon, would-be astronauts will pay a lot less and not have to go through months of rigorous training, and as a result the number of space tourists will skyrocket.

v Space tourist Dennis Tito (left) aboard the ISS with cosmonauts Talgat Musabayev and Yury Baturin.

AS YOUR EYES SURVEY THE SCENE IN FRONT YOU, IT IS IMPOSSIBLE TO DECIDE WHAT TO LOOK AT NEXT. YOUR JAW DROPS AT THE SHEER SPLENDOR OF EARTH BELOW, FILLING YOU WITH A LEVEL OF AWE AND WONDER THE LIKE OF WHICH YOU'VE NEVER EXPERIENCED BEFORE.

Lightning storms flicker and fade like flashbulbs. Sprawling metropolises—New York, London, Sydney—come and go as you drift over oceans and continents. Country borders seem more like an abstract human construction with every passing minute. Mountain ranges, archipelagos, deserts, and ice caps fill your eyes before moving on. Aurorae dance around the fringes of our world, as stars and constellations pepper the black sky beyond.

This unbelievable and enchanting panorama is just what you might expect to see through the windows of the first space hotel. Paying space tourists will only make fleeting visits to suborbital space to start with, but

∧ The kind of view you might expect from the first orbiting space hotels.

> Artist's impression of a futuristic space hotel in Earth orbit with a passenger shuttle docked at the bottom.

eventually the clamor for more time to enjoy the wonders of the cosmos will be undeniable. Just one orbit of the planet will take around ninety minutes, so in a day you'll witness sixteen sunsets and sixteen sunrises. Imagine what you might see if you spent a week circling Earth.

The concept of vacation accommodation in space is far from new. Even before the United States and the Soviet Union proved that you could build modular space stations in the 1970s, people speculated that orbiting spaceships could one day host paying customers. In 1967, Barron Hilton—son of the Hilton chain founder Conrad Hilton—spoke at the American Astronautical Society on plans for a fourteen-story space hotel large enough for twenty-four people. The Japanese architecture firm Shimizu drew up their own plans in the late 1990s, as did the Russian aerospace company Orbital Technologies in

THE BIGELOW EXPANDABLE ACTIVITY MODULE (BEAM)

April 2016 saw the first tentative steps toward a future with space hotels when NASA sent the Bigelow Aerospace Expandable Activity Module (BEAM) to the International Space Station. In a case of the old guard working with the new, it was launched on a rocket built by Elon Musk's private space company SpaceX. The man behind BEAM is hotel magnate Robert Bigelow.

Robotic arms attached BEAM to the ISS's Tranquility module. The module was designed to inflate, so that it could be packed up small and take up minimal space on the rocket. Once in orbit, it expanded to 12 by 10 feet 6 inches (3.6 x 3.2 m) when fully pressurized. It was designed to test the feasibility of creating habitable pods that could serve as rooms in a future astro-hotel. Astronauts visited the module eight times a year to see how it has coped with dangers, including wild swings in temperature, radiation, space junk, and micrometeorites. Samples from BEAM were returned to Earth for further analysis.

BEAM was set to stay attached to the ISS for two years before being jettisoned to safely burn up in Earth's atmosphere. However, in October 2017 NASA announced it would stay attached until 2020 with a possibility of further extensions.

v An artist's concept depicting the Bigelow Expandable Activity Module (BEAM) attached to the ISS.

2011. While none of these buildings have yet materialized, recent developments in inflatable space habitats may have brought space hotels one step closer (see box, The Bigelow Expandable Activity Module).

But why limit ourselves to Earth orbit? A cruise ship is effectively a floating hotel, with all the amenities you'd find in a traditional establishment. In the future, cosmic cruise liners could sail the black ocean between Earth and the Moon, orbiting our nearest neighbor for spectacular sweeps over its craters. You'll witness Earthrise before heading for home again. The whole trip could take just a week.

One of the major factors holding us back from hotels in space is that traditionally it has cost a fortune to launch objects into orbit. Yet, with the current momentum driving forward the commercialization and privatization of space, it wouldn't be surprising to find people routinely paying to live in space by the end of this century.

< NASA astronaut Kate Rubins inspecting the Bigelow Aerospace Expandable Activity Module (BEAM) in 2016.

CHARLIE AND THE GREAT GLASS ELEVATOR

Perennial children's favorite Roald Dahl was an unlikely source of an early, highly detailed account of a space hotel in fiction. In *Charlie and The Great Glass Elevator*—Dahl's 1972 sequel to *Charlie and The Chocolate Factory*—the title character accidentally travels into orbit with his family and Willy Wonka on a space elevator (see p. 132). Perhaps Dahl was inspired by NASA's Skylab crewed space station—test stages were constructed at the beginning of the 1970s and Skylab was eventually launched in 1973.

Once in orbit, Charlie and his companions are forced to dock with the unoccupied Space Hotel USA. This had been launched just two days earlier by the American government led by President Lancelot R. Gilligrass. Shaped like a sausage, Dahl's space hotel is 1,000 feet (3,280 m) long and boasts five hundred luxury bedrooms, a tennis court, a gymnasium, and a swimming pool. A gravity-making machine does away with the perils of weightlessness.

The hotel is designed so that the staff live off site in an accompanying transport capsule. Yet while working in the hotel, twenty-four staff members are eaten by an invading alien race of Vermicious Knids, who are normally prevented from attacking humans by Earth's atmosphere.

SPACE ELEVATORS

SLOWLY BUT SURELY YOUR ELECTRIC CAPSULE CLIMBS A CABLE INTO THE CLEAR BLUE SKY, LEAVING THE EQUALLY AZURE EXPANSE OF THE OPEN OCEAN BEHIND.

Gradually you accelerate upward, reaching a cruising speed of several hundred miles per hour. Even then it takes you two weeks to complete the ride beyond geosynchronous orbit to reach the top, some 62,100 miles (100,000 km) above Earth's surface. The shrinking view of the planet below more than makes up for your sedate progress.

These space elevators could be the future of space travel. After the costs of the initial build—which might top $10 billion and take 10 years of construction—the price of getting a couple of pounds into orbit could drop as low as $200. That compares to $3,000 for SpaceX's reusable rockets (see p. 14). Power would come from solar panels and lasers beamed from Earth's surface. Anything launched from the cable above geosynchronous orbit would already be traveling at greater than orbital velocity, making trips to the Moon and Mars a cinch. One spacecraft could be dispatched into the Solar System every day. However, the sensitive geopolitics surrounding such an endeavor might hinder these plans (see box, right).

The basic premise is straightforward, at least on paper. Anchor a cable to a base station out at sea on the equator and connect the other end to a 2,200-ton (2 million-kg) counterweight sitting above geosynchronous orbit. That way the counterweight stays above the base station as the planet rotates. Imagine attaching a ball to a length of string and whirling it around above your head. Just as the momentum of the ball keeps the string

^ Computer artwork of a future elevator for space transportation, with the counterweight at the top.

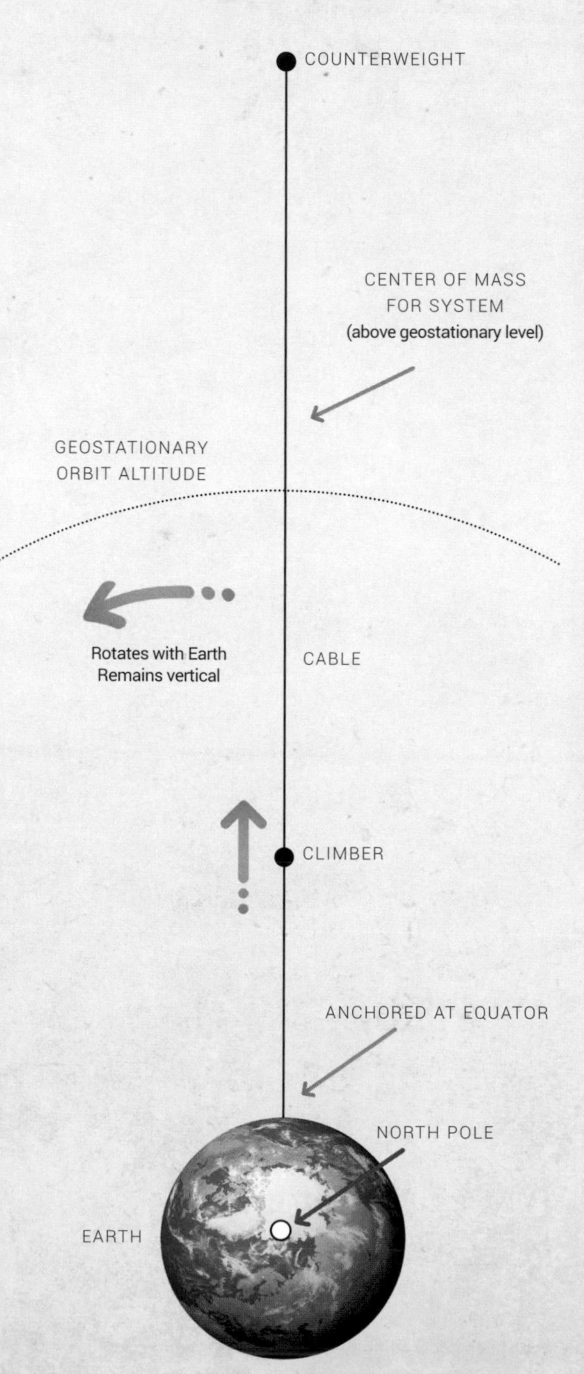

THE GEOPOLITICS OF SPACE ELEVATORS

Such a gargantuan engineering project is probably beyond the means of any one sovereign nation and would require unprecedented international cooperation. That's easier said than done on a planet that saw two world wars in the previous century and whose inhabitants are increasingly nervous of a nuclear conflict in this one. However, the International Space Station is a beacon for what can be achieved through peaceful cooperation in space. Perhaps it will have to be overseen by the United Nations. Even then, individual countries might get nervous about the use of the elevator for spying.

The history of the Suez Canal perfectly illustrates the issues with having crucial global transportation infrastructure in just one country. So, not only would it need to be far from existing commercial aircraft routes, but a space elevator would have to be in international waters (probably in the middle of the Pacific Ocean). It would also be a tempting target for terrorists and rogue nations and therefore it would need to be heavily defended with permanent military surveillance. Fighter jets and submarines might patrol the borders of the no-fly zone. Monitoring the area from space would be essential. In many ways, these political obstacles may prove to be more difficult to overcome than the engineering challenges.

taut, so the motion of the counterweight provides the necessary tension in the space elevator cable.

The cable would stretch to 62,100 miles (100,000 km), but would be only about 40 inches (1 m) wide—100 million times longer than across. That's the same ratio as a human hair more than 6 miles (10 km) long. It would have to be made of a material weighing no more than about 5 ounces (150 g) per 40 inches (1 m). Finding such a lightweight material that's also strong enough is a major challenge. Possible options include carbon nanotubes and graphene (see box, p. 134). The cable would be built using 22-ton (20,000-kg) climbers, but once finished it might be able to carry as much as 1,100 tons (1,000,000 kg) at a time.

< A schematic of a space elevator anchored to Earth at the equator and the counterweight above geostationary orbit.

Several options have already been explored for the counterweight. Millions of pounds are required, which rules out launching it from Earth. Perhaps we could tow an asteroid in from deep space, although that could be potentially hazardous if misjudged. Alternatively, we could gather up existing space junk, potentially solving two problems at once (see pp. 108–11). Other suggestions have included fashioning the counterweight from the spent climbers used to construct the cable.

> Graphene—a layer of carbon just one atom thick—could be used to fashion a space elevator cable.

For years, space elevators seemed more than a little farfetched, consigned to the pages of science-fiction stories, such as Arthur C. Clarke's *The Fountain of Paradise*. The reason is that we've never had a strong enough material from which to build the main cable. That may have changed, thanks to the discovery of graphene in 2004 by a team of physicists working at the University of Manchester in England. Made of a layer of carbon just one atom thick, graphene is remarkably strong for its light weight. The fact they found it by accident, discovering new things like children who are free to play and learn, is a testament to the power of "blue skies" scientific research.

Graphene has potential applications across many sectors and has been lauded by many people as a "wonder material." It may also help get us to space more easily one day. In 2017, a team of researchers from the Massachusetts Institute of Technology (MIT) used two-dimensional sheets of graphene to fashion a sturdy three-dimensional structure. It's 10 times stronger than steel, but only 5 percent as dense. Engineers are looking into using it as a replacement for steel in conventional architecture, but the researchers have also noted its potential application to the space elevators of the future.

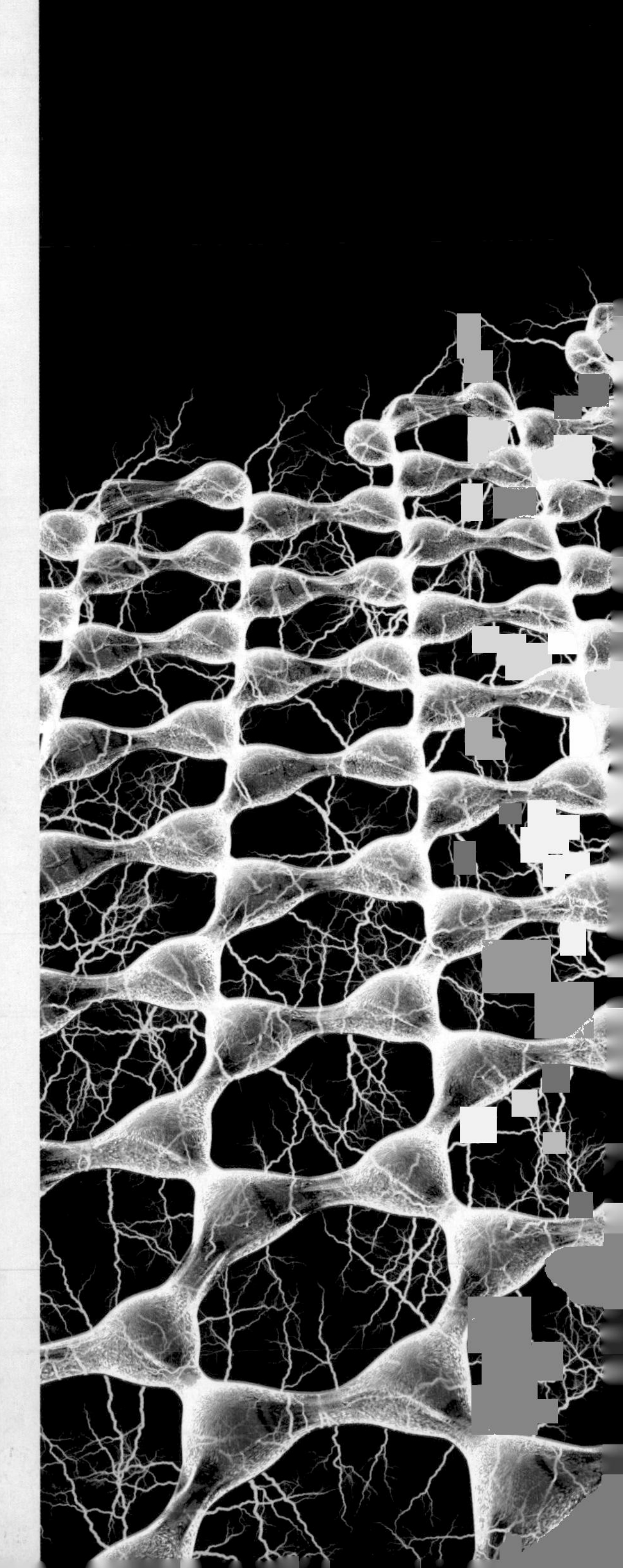

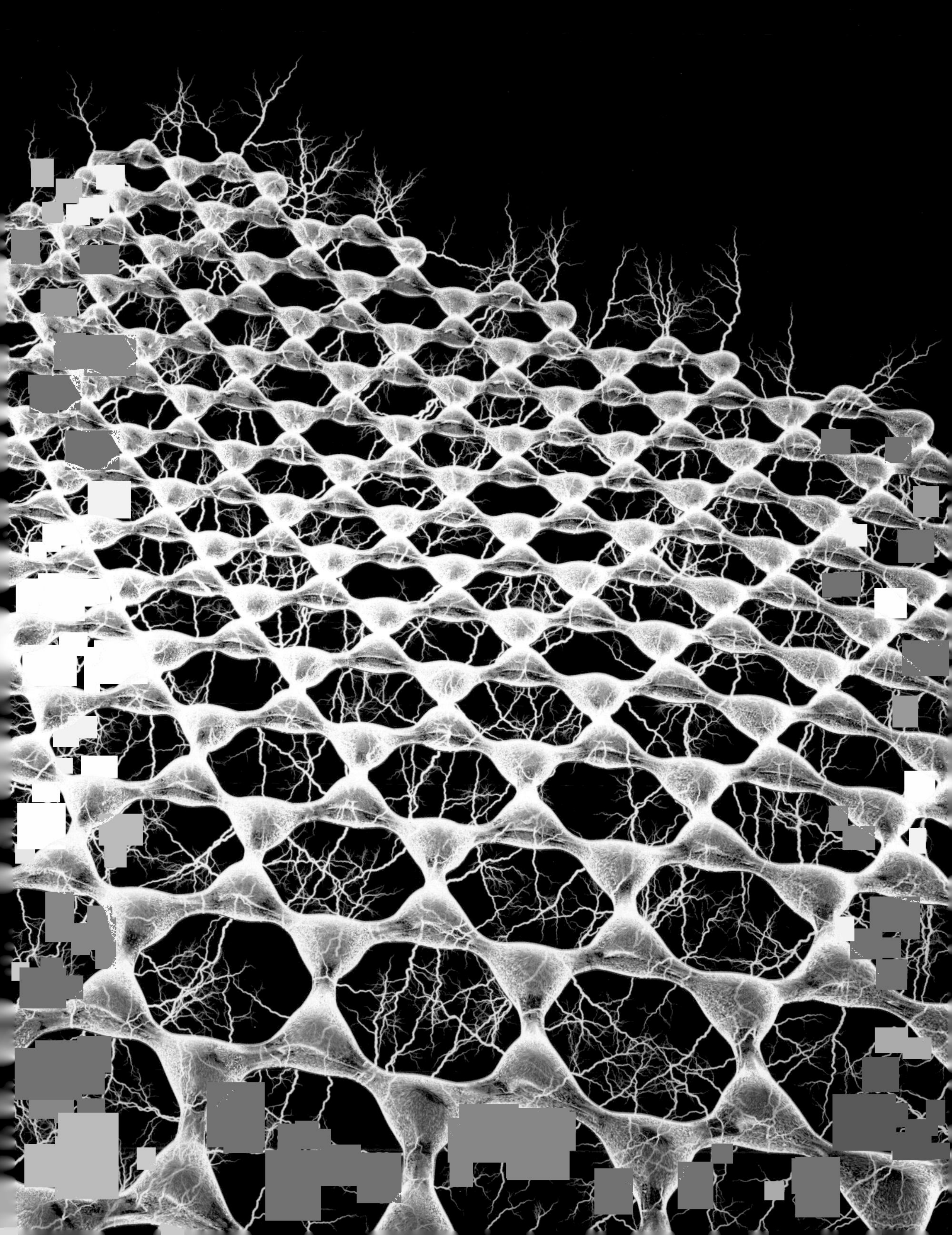

ON CHRISTMAS EVE 1968, APOLLO 8 ASTRONAUT BILL ANDERS TOOK OUT HIS CAMERA DURING THE FIRST CREWED MISSION TO ORBIT THE MOON. IN HIS VIEWFINDER WAS THE EARTH RISING ABOVE THE LUNAR SURFACE.

It is one of the most iconic images in the history of space exploration. "Oh my God! Look at that picture over there! There's the Earth coming up. Wow, is that pretty," he said to his crew-mates Jim Lovell and Frank Borman. That day they sent a Christmas message back to Earth—NASA had told them that they would have the largest audience ever to listen to a human voice.

Apollo 8 made 10 laps of the Moon, each lasting about two hours. They got within 68 miles (110 km) of the lunar surface. Our planet appeared almost cut in half, the dark line of nightfall straddling Africa. Swirling clouds twisted across the sunlit surface, with Earth's deep blue oceans bright beacons against the inky black space beyond.

Just as we see the Moon wax and wane through a series of phases over a period lasting nearly a month, future inhabitants of lunar space stations and Moon bases

> Earth rising above the lunar surface, as seen by the astronauts of Apollo 8 on Christmas Eve 1968.

ZOND 5

Frank Borman and his Apollo 8 colleagues might have been the first humans to orbit the Moon, but they weren't the first living things to do so. That honor goes to the animal crew of the Soviet Zond 5 mission, who beat the Americans to lunar orbit by a few months. On board were two tortoises, mealworms, wine flies, bacteria, plants, and seeds. A life-size mannequin fitted with radiation sensors sat in the pilot seat to probe the risk to future human astronauts.

Zond 5 flew within 1,240 miles (2,000 km) of the lunar surface before being catapulted back to Earth by our satellite's gravity. The spacecraft crashed down into the Indian Ocean on its return instead of landing in Kazakhstan as planned. Nevertheless, all aboard were relatively healthy. The tortoises had lost 10 percent of their body weight, but that was the worst of it.

The success of the mission spooked the United States into accelerating the Apollo program in fear of being beaten to the Moon by the Soviets. Originally, Apollo 8 had been scheduled for a test run in high Earth orbit, but its destination was swiftly switched to the Moon.

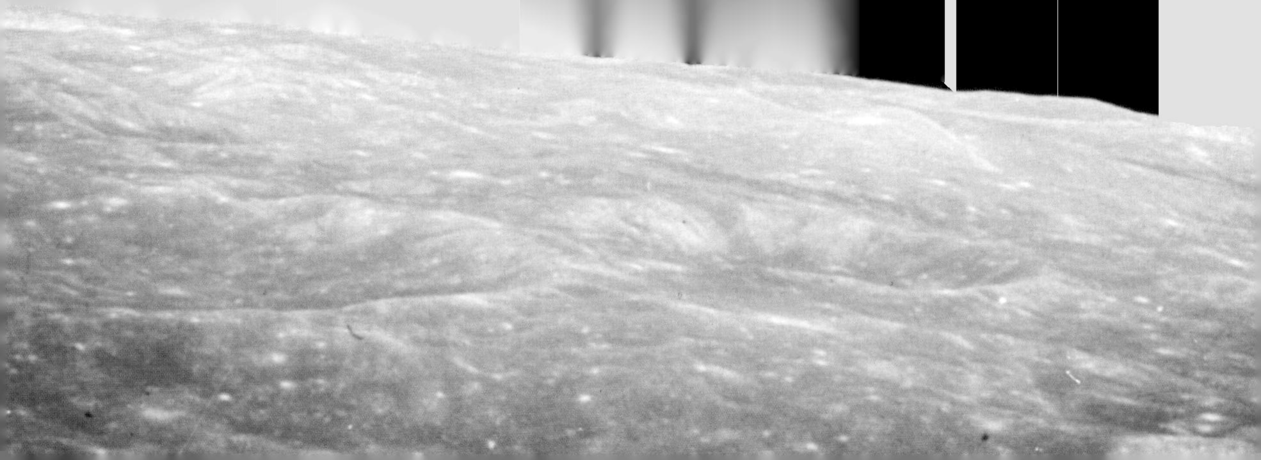

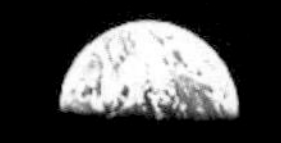

will see Earth appear to transform from a tiny crescent to being full over the same period. None of the Apollo missions stayed on the Moon long enough to see a full cycle of Earth phases. The longest mission—Apollo 17—was also the last.

Only six men—the Apollo Command Module pilots—know what it feels like to be completely cut off from Earth and every other human being. Michael Collins, Richard Gordon, Stuart Roosa, Al Worden, Ken Mattingley, and Ronald Evans all stayed in lunar orbit while their colleagues descended to the lunar surface. For forty-five minutes of every orbit they experienced a communications blackout, with the Moon itself blocking any radio link back to Earth or the other astronauts. Any future occupants of a lunar space station (see box, below) would have to contend with a similar situation, although the company of their fellow crew-members should make the isolation feel less acute.

What might a future human crew get up to? One idea is to park a spacecraft at a point where the gravity of Earth and the Moon balance so that astronauts can drive rovers across the lunar surface in real time. It would be an excellent trial run for doing the same thing one day on Mars.

A return to lunar orbit might come sooner rather than later, too. SpaceX has announced its intention to send two paying private citizens around the Moon in the coming years, perhaps ushering in a new era in our long love affair with our nearest neighbor.

< *Apollo 11* astronaut Michael Collins's view of Armstrong and Aldrin returning from the lunar surface.

> An artist's impression of a future space station in lunar orbit.

In September 2017, NASA and the Russian space agency Roscosmos released a joint statement at the International Aeronautics Congress in Adelaide, Australia. They publicly declared their intention to team up and build a space station in lunar orbit: the Lunar Orbital Platform-Gateway (LOP-G). Construction is scheduled to begin in the 2020s, with the finer details currently being worked out through a series of smaller concept projects. Other countries are being consulted, too, in the hope of making it an International Lunar Station. The plan is to have a crew of astronauts spending a year orbiting the Moon by the end of the 2020s. The ease of access to the surface would surely see human footprints in the lunar dust once again.

The mission would extend our long-duration space flight capabilities and test new technologies significantly farther from Earth than the International Space Station. The astronauts would still be able to make it to the safety of home within a few days if a crisis led to the mission being abandoned. Equally, an emergency supply ship could rendezvous with the station relatively quickly. It's a natural stepping stone on the path to sending people to Mars for the first time later this century.

MANY CONSIDER IT THE GREATEST ACHIEVEMENT IN HUMAN HISTORY. BETWEEN 1969 AND 1972, IN A FLURRY OF ACTIVITY, TWELVE AMERICAN ASTRONAUTS LEFT THEIR FOOTPRINTS IN THE LUNAR DUST.

The footprints on the Moon still exist, because there's no wind, rain, or water to erode them away.

The Apollo era was the response to President John F. Kennedy's bold call to land on the Moon before the end of the 1960s. It was a way to outmaneuver the Soviet Union during the Cold War. That the Soviets remained quiet, and even sent a robotic probe to try to get the first Moon rocks back to Earth ahead of Apollo 11, should be all the evidence needed to quell the conspiracy theorists who say it never happened.

Each mission grew increasingly ambitious, eventually exploring more of the Moon using multimillion-dollar lunar buggies to drive across the lunar surface. The Apollo dozen spent a total of eighty hours exploring the Moon and returned with nearly 880 pounds (400 kg) of rocks for analysis. Alan Shepard played golf. Dave Scott dropped a hammer and a feather to show that objects of different masses fall at the same rate in a vacuum. More important, they inspired a generation of scientists and engineers to push the envelope of what's possible.

Those developments have led us to the era of the commercialization of space and the real possibility of people returning to the Moon in the not-too-distant future. In October 2017, Vice President Mike Pence directed NASA

> Buzz Aldrin was photographed on the Moon by Neil Armstrong, who can be seen in the reflection on the visor.

THE GOOGLE LUNAR XPRIZE

Since the Apollo 17 astronauts set a course for home in 1972, no human has returned to the Moon. To make matters worse, only three robotic probes have landed on the surface of our nearest neighbor in all that time: the Soviet Union's Lunokhod 2 rover in 1973, Luna 24 in 1976, and the Chinese Yutu rover in 2013. Other missions have orbited around the Moon and deliberately crashed probes into the lunar surface, but there have been no other soft landings.

So far, all missions to safely land on the Moon have been sent by space agencies backed by national governments. The Google Lunar XPRIZE aimed to change that by offering $20 million to any private outfit that could land a rover on the Moon, drive 1,640 feet (500 m) across its surface, and send back high-definition video and images. The prize for second place was $5 million. Five teams—SpaceIL (Israel), Moon Express (US), Synergy Moon (International), TeamIndus (India), and HAKUTO (Japan)—secured contracts to get their spacecraft to the Moon. However, Google announced in January 2018 that the race would end without a winner. It remains to be seen how long it will take until a commercial venture makes the journey.

∧ The next flags and footprints on the Moon could well be Chinese.

> The far side of the Moon, as seen by the astronauts of Apollo 16.

to start planning for sending humans back to the Moon. However, the next footprints in the lunar dust may be left by Chinese feet. The Deputy Director General of the China Manned Space Agency announced in 2017 that they are making "preliminary preparations" for a Moon shot.

The successes of the Apollo missions may give rise to a sense of complacency about the ease of landing on the Moon, however, there have been casualties and crises along the way. The Apollo 1 astronauts died during an explosion on the launch pad. The famous lucky escape of Apollo 13 is another reminder that it's not all plain sailing. Even during Apollo 11, Neil Armstrong had to take manual control and guide the Eagle lander over a sizable field of boulders with fuel ebbing away by the second. Michael Collins, the Commander Module Pilot left to orbit the Moon, said: "My secret terror for the last six months has been leaving them on the Moon and returning to Earth alone; now I am within minutes of finding out the truth of the matter."

A successful Moon landing requires a lot of things to go right, one after the other. Collins called it a "fragile daisy chain of events." We may have to tolerate failure before success becomes the norm once again.

Contrary to popular belief, there is no permanently dark side of the Moon. Half is always illuminated and the other half is in shadow, but these areas constantly shift as the Moon orbits us. The Moon does, however, have a far side. Our satellite is tidally locked to Earth, meaning that it always presents us with the same face. Over time, the Moon's rotation period slowed down until it matched the time of one lunar orbit: 27.3 days.

Consequently, the far side of the Moon represents a largely unexplored treasure trove. All Apollo missions have landed on the near side. We've sent lunar orbiters around the back for a look, but landing a human mission there would see our understanding escalate, especially if we could return rock samples. The Chinese space agency is already intending to send the Chang'e 4 robotic rover there at the end of 2018. A separate probe is needed to relay signals from Chang'e 4 back to Earth, because the direct path for signals is blocked by the Moon itself. That would make it an ideal spot for a radio telescope—with the Moon handily obscuring noise from Earth in return.

YOU WAKE UP, STRETCH, AND ROLL OUT OF BED. WALKING OVER TO THE WINDOW YOU DRAW BACK THE CURTAINS, ONLY TO WAVE AT EARTH AS OUR BLUE MARBLE PLANET CLIMBS HIGH ABOVE THE ASHEN HORIZON.

You then bounce down the hall to tend to the microgravity vegetable garden. This unusual start to your day is because you're one of the first inhabitants of Lunar Village 1—humankind's first permanent dwelling on another world.

This might sound farfetched, but if everything goes to plan, it could become reality a lot sooner than you think. Space agencies around the world, including NASA, the European Space Agency (ESA), and Roscosmos (the Russian space agency), are already looking into the possibility. The head of ESA is pushing for an established human presence on the Moon by the 2030s. A spokesperson for the Chinese space agency has revealed they've had conversations with ESA about a possible collaboration. Excitingly, recent technological developments have put such a lofty goal within touching distance (see box, p. 145).

A key question for any would-be lunar citizen is where to set up camp. Compared to Earth, the Moon is a hostile, alien environment—no liquid water, no atmosphere, and unusual periods of day and night. The consensus of opinion seems to be to head for the lunar south pole. There are several reasons for this. First, one rotation of the Moon on its axis takes nearly a month, meaning that most of the lunar surface is bathed in daylight for two straight weeks before being plunged into enduring darkness for the following two weeks. Go for the south pole, however,

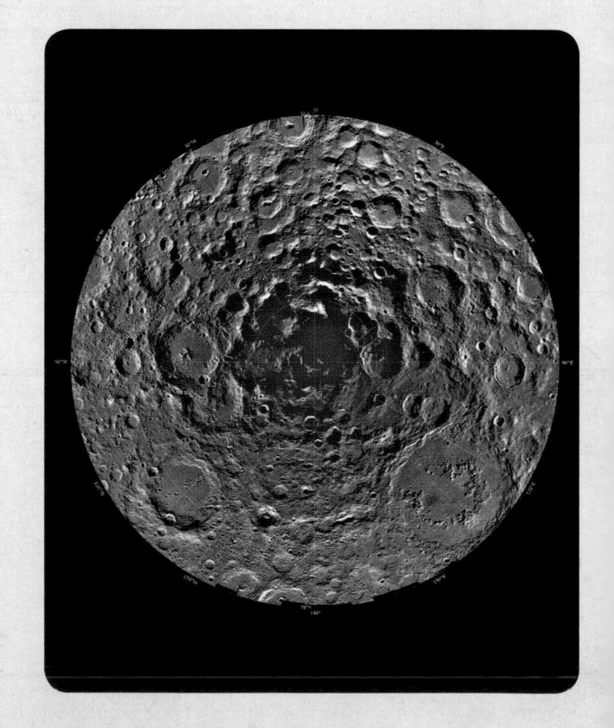

^ Map depicting craters on the south pole of the Moon.

> Architects Foster+Partners have joined with the European Space Agency to test 3D printing using lunar soil.

3D PRINTING ON THE MOON

Using lunar materials efficiently is key to a successful lunar base and the survival of its inhabitants. Space scientists call it ISRU, which stands for "In Situ Resource Utilization." The main benefit is in saving money; at present, it costs at least $3,000 to launch a single kilogram (about 2¼ pounds) of material into Earth orbit, let alone to a lunar outpost. For this reason, the three-dimensional (3D) printing revolution is causing a stir in rocket science circles. In 2014, a wrench was printed on board the International Space Station, its design emailed up from Earth. It is much cheaper than making it on Earth and sending it on a rocket.

So researchers have been investigating whether future Moon builders could feed a 3D printer with lunar soil and create the material from which to construct a home. Piling the soil on top of the base would help shield inhabitants from the dangers of space radiation. 3D printing also has the benefit of getting help to the Moon more quickly in case of an emergency. Imagine that a key part of the base underwent catastrophic failure. Mission Control on Earth could beam them the design for a 3D-printed replacement part in seconds instead of them havnig to wait for days while it is sent by spaceship.

∧ Barry "Butch" Wilmore shows off a ratchet wrench made with a 3D printer on the ISS.

and the tilt of the Moon means you're in sunlight 90 percent of the time. That solar energy is a precious commodity and would be gobbled up by the banks of solar panels we'd build to power our Moon colony. The 13-mile (21-km) Shackleton crater, named after Antarctic explorer Ernest Shackleton, is top of the hit list.

We also know from our robotic exploration of the Moon that its south pole is home to significant quantities of water ice. Some southern craters are so deep that their bases are never kissed by sunlight, ensuring that the ice persists. Melting the ice would give us water to drink and clean with, but it is important in other ways, too. A water molecule's chemical structure—H_2O—comprises two atoms of hydrogen and one of oxygen. If you could pry these constituents apart, you would not only have oxygen to breathe but hydrogen to convert into rocket fuel. Splitting water apart could be done using energy derived from the solar panels.

> European Space Agency illustration of a multi-dome base being constructed on the Moon using 3D printing.

Future lunar colonists might use the Moon's resources for more than just their own survival. Inhabitants of the first permanent English settlement in North America—Jamestown, Virginia—harvested tobacco and shipped it home for profit. Moon dwellers may also mine the lunar environment and send precious commodities back to Earth. Our natural satellite is thought to have substantial quantities of a rare form of helium called helium-3, generated by interactions with the solar wind. It's a crucial ingredient in some of the nuclear fusion reactions that, in the future, power stations might use to generate clean and plentiful electricity. Harrison Schmidt—the geologist who flew to the Moon aboard Apollo 17—has long backed the idea of mining the Moon for helium-3.

Moon Express is a private company granted unprecedented permission by the US government to fly beyond low Earth orbit. As of 2017, they had raised almost $50 million in private investments with the intention of sending a technology demonstration mission to the Moon. They refer to the Moon as Earth's eighth continent—it has a total land mass roughly equivalent to North and South America combined. And they've made no secret of their desire to set up a trade route between it and Earth's traditional seven continents.

esa

IMAGINE SEEING EVERYTHING YOU KNOW AND EVERYONE YOU LOVE GRADUALLY RECEDING, AS OUR BLUE PLANET SHRINKS TO A DISTANT DOT.

Could you cope? Mars travelers would need to be made of strong stuff. They'd need to be calm under pressure; comfortable with prolonged bouts of isolation; good at building interpersonal relationships and conflict resolution; and focused on the task at hand. And, there's no getting away from it, they'd need to be prepared to offer their life to the cause of pushing the limits of human endeavor.

"I think the first journeys to Mars are going to be really very dangerous. The risk of fatality will be high; there's just no way around it," SpaceX founder Elon Musk told the International Astronautical Congress in 2016.

Scientists are already preparing for the first human trip to Mars by simulating the psychological rigors of the mission as closely as possible (see box, right). They're also considering the best mix of genders (see box, p. 150). A variety of ages would be good, too. A blend of youth and experience would bring a range of knowledge and ingenuity to problem-solving. Some have argued that we should send only those who've finished having children, or those who don't want to have kids, because of the threat of sterility from the harsh radiation exposure during the trip.

> An astronaut training on the simulated Martian terrain that formed part of the Mars-500 project.

↗ The crew of the Mars-500 project measuring their brain activity using an electroencephalogram (EEG).

Both homesickness and increased anxiety are already known problems among existing astronauts, and they can still see Earth up close, with much of the familiarity of home. The spectacular view is enough to mitigate some of the negatives. Halfway through the journey to Mars, however, and neither your home planet nor your destination world would be seen in any detail. Long-term confinement in a small space has been shown to prompt just as many psychological issues as social isolation. With Earth so far away, a trained psychologist seems a good bet in any crew. That, or a sophisticated artificial intelligence-based interface offering meaningful immediate remedies, such as

THE MARS-500 PROJECT

Between 2007 and 2011, Russia, China, and the European Space Agency (ESA) ran several experiments to simulate the psychological effects of a human trip to Mars. During one mock-up, an all-male crew of six spent 520 days holed up inside a test facility in Moscow. That's the same amount of time as a return trip to the Red Planet plus 30 days on the Martian surface. The landing was simulated, along with three "Mars walks," all within the confines of a space totaling just 19,423 cubic feet (550 cu. m). Contact with the outside world was restricted and came with a built-in twenty-minute delay—the typical amount of time it would take a radio message to travel from Mars to Earth and back again.

The lack of natural sunlight caused the crew's sleep patterns to drift out of rhythm. Some were asleep when others were awake, hampering team efforts. Sleep deprivation led some

∧ The crew of the Mars-500 project during a press conference.

to perform poorly on computer tasks. Their energy levels also dropped significantly on the return journey, once the excitement of "Mars" had worn off. They spent an extra seven hundred hours in bed compared to the outward journey. Boredom, it seems, was one of the biggest problems they faced. All of these issues should be considered when selecting the right crew.

cognitive behavioral therapy (CBT), a well known talk-based treatment for mental health problems.

Genetic research might play a part in crew selection, too. We know that certain genetic differences cause some astronauts to experience vision problems that others do not. It's not hard to imagine a future where would-be astronauts are genetically screened in the selection process and ruled out based on having a greater risk of being affected by radiation or a propensity to psychological problems.

All of these reasons have led some to argue that, before we head to Mars, we need to go back to the Moon to learn the lessons of being far from Earth for prolonged periods without all of the risk.

> The crew of the *Daedalus* in the National Geographic TV show *Mars*.

The only humans to have left low Earth orbit (LEO) so far have all been male. However, there's a growing body of research suggesting that this isn't the way to go when selecting the optimum crew for the first trip to Mars. One of the best comparable situations we have is in expeditions to Antarctica, because polar explorers are also isolated in extreme conditions in an unfamiliar environment. A 2004 study looked at mixed-gender groups at the French Dumont d'Urville Antarctica Station. Adding women to the crew saw a positive improvement in the group dynamic by making the men less rude, although things inevitably became complicated when the men and women were of a similar age—love interests and rivalries caused tension in the group.

Sheryl Bishop, a scientist who has worked on the Mars Desert Research Station—a Red Planet analog experiment in Utah—believes that an all-female crew isn't ideal, either. She's found women tend to focus too much on interpersonal relationships. This is supported by the Chief Medical Officer of the Mars One project—Norbert Kraft—who found that a crew with a 50:50 mix of genders performed better than groups of one gender alone.

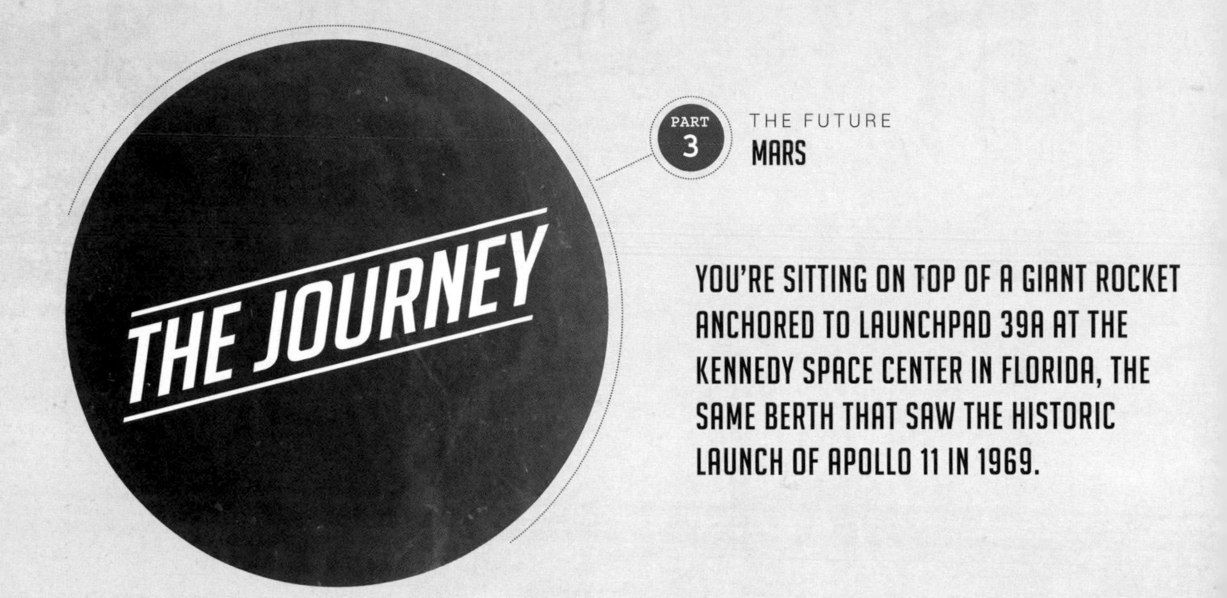

PART 3 THE FUTURE
MARS

YOU'RE SITTING ON TOP OF A GIANT ROCKET ANCHORED TO LAUNCHPAD 39A AT THE KENNEDY SPACE CENTER IN FLORIDA, THE SAME BERTH THAT SAW THE HISTORIC LAUNCH OF APOLLO 11 IN 1969.

You're about to make history of your own by leading the first human expedition to another world. With a good slice of fortune, you'll walk out on the surface of Mars seven months from now. You'll have plenty of time to write your own "It's one small step" speech. What will you say?

The Red Planet has beguiled us for centuries, teasing us with clues that it might once have been like Earth and offering the tantalizing prospect of a second home for humanity. We've already sent an armada of robotic probes to learn more about its secrets. Humans will soon follow, in all likelihood by the middle of this century. It won't be easy, landing small rovers on Mars is hard enough (see box, p.159).

Of those claiming to be heading for the Red Planet in the decades ahead, perhaps the most thought-through plan belongs to SpaceX, the private space company owned by entrepreneur Elon Musk. He's already revolutionized access to low Earth orbit and has expressed a desire to die on Mars. Musk's plan is to establish the first human colony on the Red Planet. He hopes people will sell all their earthly possessions in return for a one-way ticket. If the exodus from Europe to the New World in the last millennium is anything to go by, there will be people willing to gamble in the hope of a new beginning.

You'll fly on the so-called BFR spaceship, at about 160 feet (48 m) tall and 30 feet (9 m) across, including the propellant tanks and engines behind the living quarters.

LAUNCH WINDOWS

You can't just launch to Mars at any time you like. At least not if you want to do it economically. Unlike the Moon, Mars's distance from Earth varies considerably because both planets orbit the Sun at different speeds. At their closest, the two planets are 34 million miles (55 million km) apart. Yet they can be separated by up to 249 million miles (401 million km). So heading for the Red Planet when the trip is the shortest makes a real difference. This happens during so called "launch windows," which occur every 780 Earth days. So, in the future, an armada of spaceships might gather in low Earth orbit awaiting the optimum opportunity to blast en masse toward Mars.

The European Space Agency's robotic ExoMars mission will take advantage of the launch window in the summer of 2020. If SpaceX get their way, they could send a cargo mission to Mars during the 2022 window. Elon Musk hopes humans will follow during the next available window in 2024, but delays could see them travel in 2026 or 2029 instead. NASA might go in the 2030s and could choose 2031, 2033, 2035, 2037, or 2039, although the Trump administration has focused more on a return to the Moon.

v An artist's impression of a SpaceX crewed spacecraft approaching Mars.

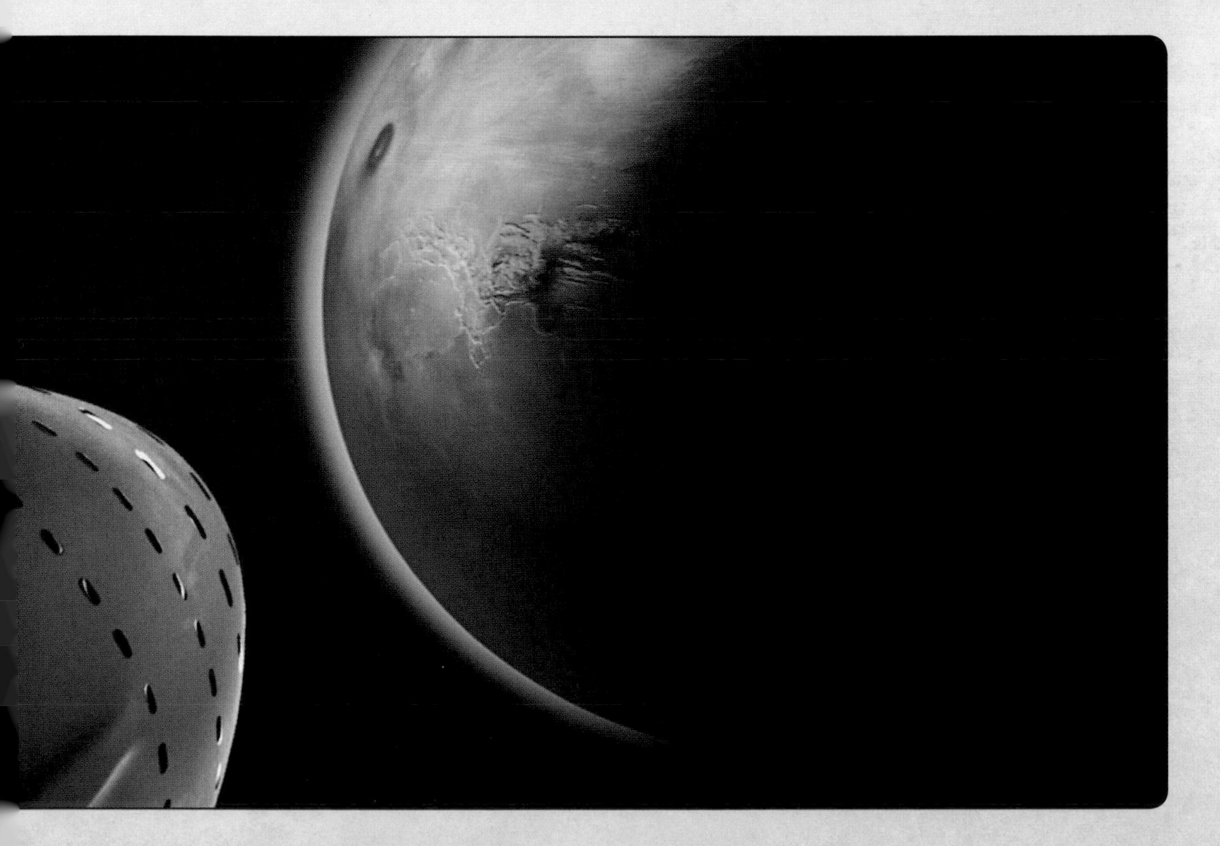

Until recently, a human mission to Mars seemed like a pipe dream—a whimsical notion several centuries in the distance. We simply didn't have a big enough rocket to launch a heavy payload with enough speed to escape Earth's gravity and head for the Red Planet. That changed in February 2018, when Elon Musk's company SpaceX successfully tested their Falcon Heavy rocket, immediately making it the most powerful launch vehicle around.

Ever the showman, and with a sharp eye for making a media splash, Musk put his own Tesla sports car on top of the rocket and sent it way beyond our planet, out past Mars and toward the asteroid belt. For SpaceX, it is just the start. With the Falcon Heavy technology successfully demonstrated, all their attention will now turn to an even bigger rocket—the BFR. Musk expects a full test of that technology within three to four years. He has a history of setting ambitious deadlines and missing them. Yet his track record so far shows that he gets the job done. Soon we may have all the hardware we'll need to start contemplating the first human trip to another planet. Mars, it seems, is now a question of when, not if.

The area for the crew of one hundred has a volume of 29,135 cubic feet (825 cu. m), larger than the cabin of the world's largest commercial passenger plane—the Airbus A380. You'll be rocketed to Mars by six engines powered by 1,200 tons (1,100,000 kg) of propellant inside a carbon-fiber fuel tank. That'll help you to reach a top cruising speed of 62,000 miles (100,000 km) per hour and reach Mars in just over two hundred days.

Life on board is pretty intimate, with only forty cabins. The design includes some large communal areas, a galley, and a solar storm shelter. Exercise is as crucial as on the International Space Station (see p. 72), so there'll be equipment for keeping fit. Food will be pretty dull, because fresh produce and space are both at a premium. High-calorie nutrient bars might be a staple. The crew will also have to contend with strains on their mental health (see p.148), but then making history never comes easy.

< SpaceX's Falcon Heavy rocket during its historic launch from the Kennedy Space Center in February 2018.

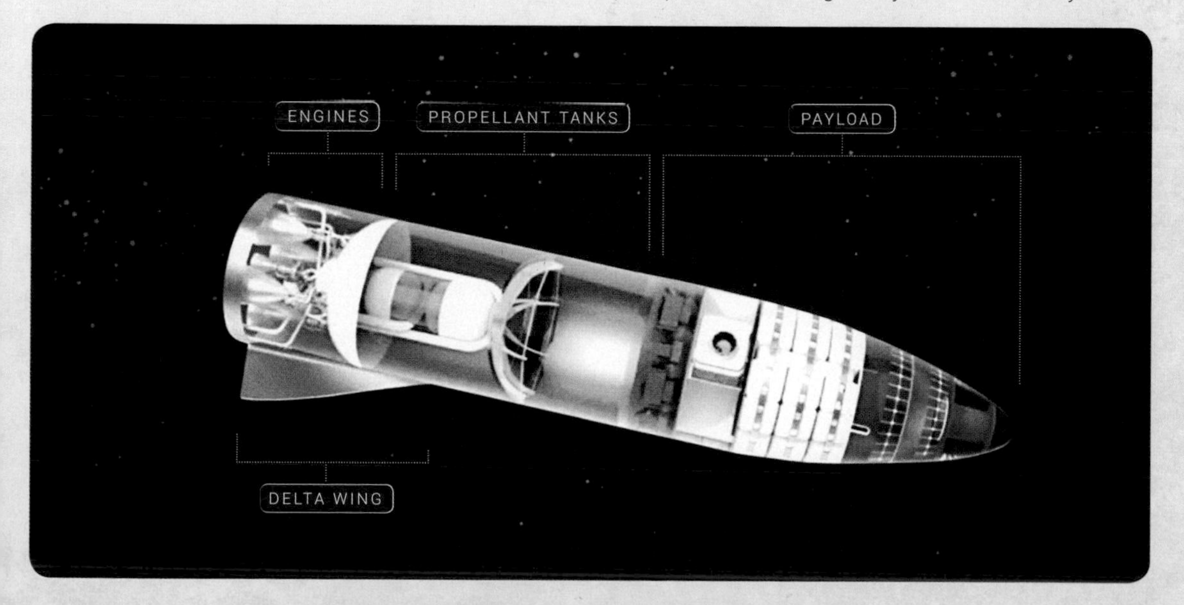

v A schematic of the SpaceX BFR spaceship that may one day carry people to Mars.

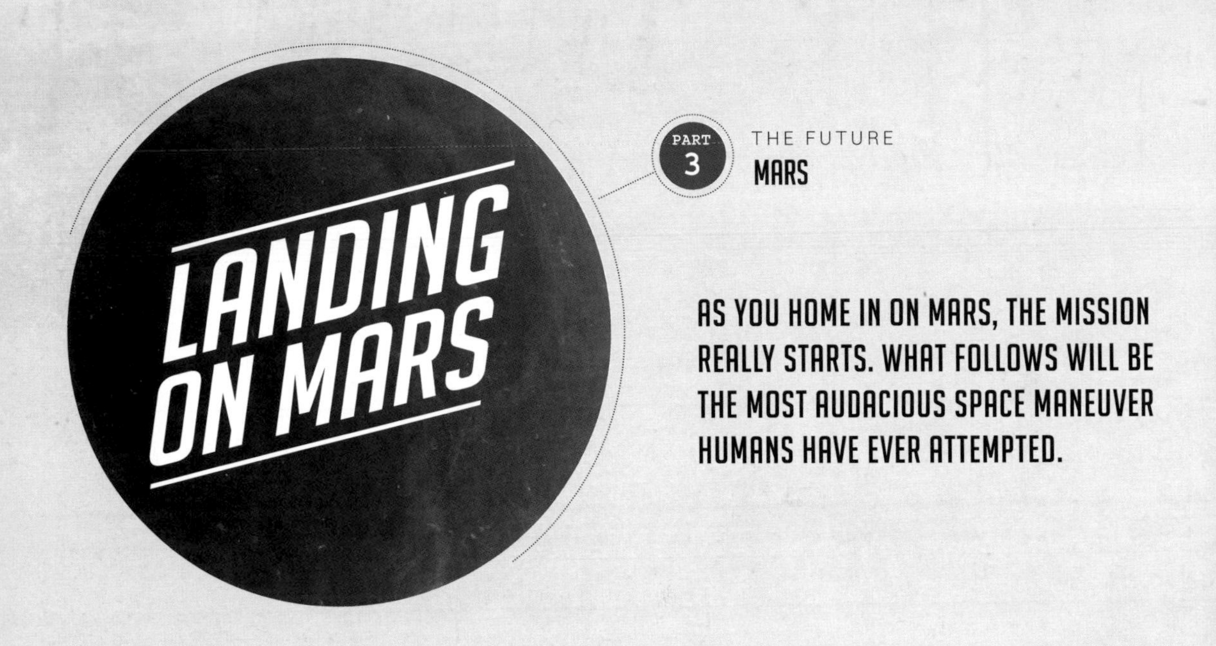

AS YOU HOME IN ON MARS, THE MISSION REALLY STARTS. WHAT FOLLOWS WILL BE THE MOST AUDACIOUS SPACE MANEUVER HUMANS HAVE EVER ATTEMPTED.

Although smaller than Earth, Mars is still a sizable planet. Its gravity is enough to pull you in hard, but there isn't much in the way of atmosphere to slow you down. Astronauts landing on Earth use our planet's gas layers as a brake, but the Martian atmosphere is only 1 percent as thick. What's more, its density changes throughout the year and across the planet due to seasonal variations in temperature and dry ice. So you might spend some time orbiting around Mars scoping out the best spot to land from a selection of predetermined choices (see box, right).

Preparations for landing will have already begun 45 days out from Mars to make sure you hit the correct approach trajectory. Mission scientists behind the rovers that have already been sent to the Red Planet dub the crucial last phase as "the seven minutes of terror." As you hit the Martian atmosphere, temperatures on the hull of the ship rise to 3,092 degrees Fahrenheit (1,700 degrees Celsius). This friction will slow you a little, but SpaceX plans to fire retrorockets at just the right time to slow you enough to land softly on the surface. NASA has looked into using a parachute, but it would have to be more than 100 feet (30 m) in diameter—twice as big as the one used to land the Curiosity rover on Mars in 2012 (see box, p. 159).

You have to get it right. One wrong move and the crew is lost. More than half of our robotic missions to Mars have failed in some way. You're all on your own, too. Due to the time delay in sending signals between the planets, when Mission Control gets word you've reached the top of the Martian atmosphere, you will already be on the ground (or not). They cannot help you.

Landing is only half the battle. To return home, you'll need to launch from the surface of Mars. That's where the Martian gravity that helpfully pulled you in turns on you. The planet's escape velocity is 3 miles (5 km) per second. That's going to require a huge amount of rocket fuel to reach. Take it along from the outset, and it adds significant weight to your mission and makes the launch from Earth and the initial Mars landing all the more difficult. The alternative is to use Mars's natural resources to manufacture the necessary propellant in situ. Yet this is far from easy, and if it cannot be achieved, you face the real possibility of spending the rest of your days marooned on Mars. Would you be prepared to take that risk?

WHERE EXACTLY SHOULD WE LAND?

Just as with the Apollo missions to the Moon, selecting the best landing site for a human Mars trip is crucial. Mission scientists want to strike a balance between a safe touchdown and settling close to places with the most natural resources and scientific interest. In October 2015, nearly two hundred scientists attended NASA's first Landing Site Workshop for Human Missions to the Surface of Mars at the Lunar and Planetary Institute in Houston, Texas. There were forty-seven different options discussed, with contenders across pretty much all of the Martian surface. However, the majority were away from the poles and at elevations below Mars's equivalent of sea level. That would ensure warmer temperatures and a higher atmospheric pressure. The next stage is to chart the subsurface ice deposits around the most promising landing sites. That mapping exercise was given the green light in November 2017.

One of the standout landing sites is the Southern Meridiani Planum—the same Martian plain explored by NASA's Opportunity rover. A human mission to the region could provide an invaluable insight into the history of the area, including its past relationship with liquid water and the chances of past life there.

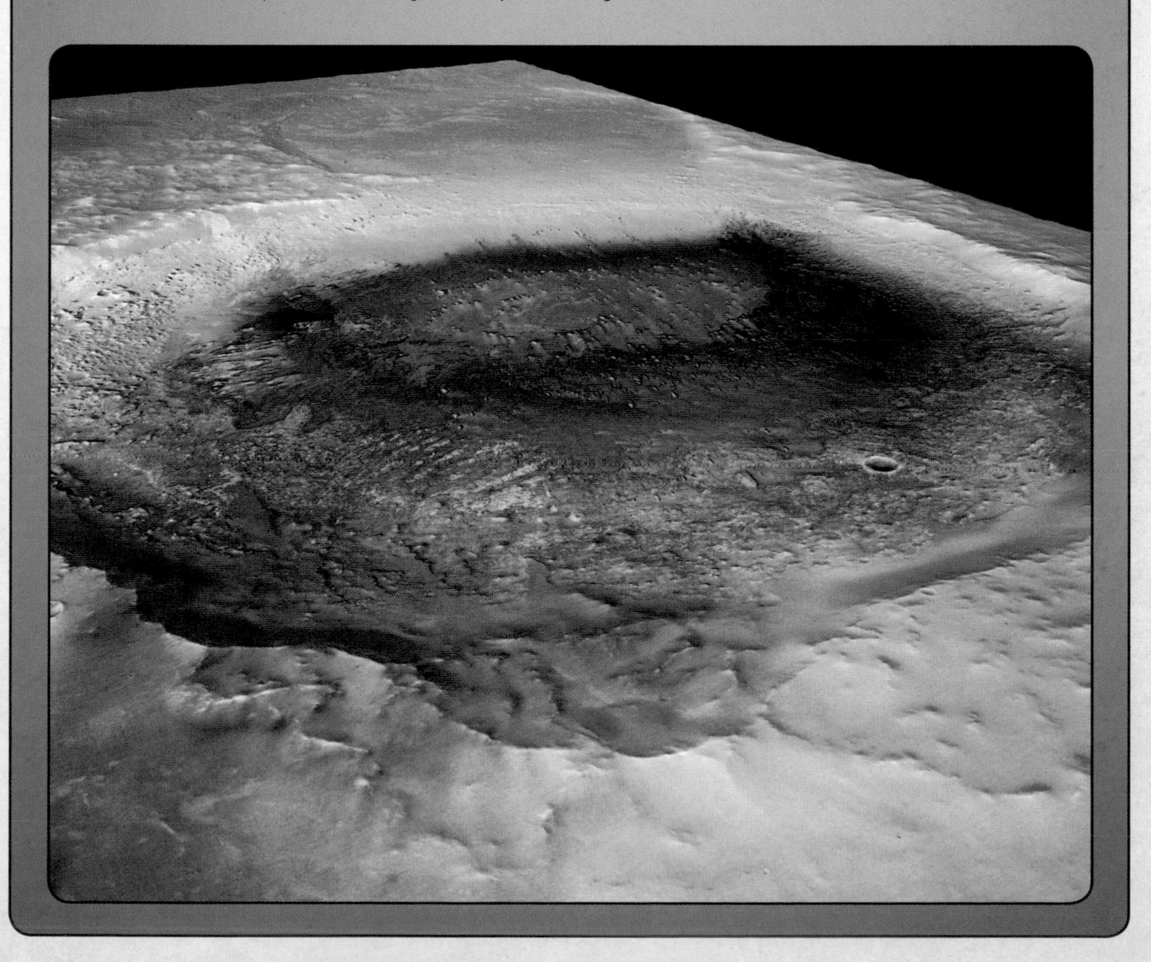

v The Meridiani Planum at the northern edge of the southern highlands of Mars.

↘ An artist's impression of a SpaceX craft landing on the surface of Mars.

Landing on Mars is difficult, even for relatively small payloads such as the Mars rovers. Approximately 5 feet (1.5 m) tall and wide, both Spirit and Opportunity were dropped onto the Martian surface inside inflatable airbags. This cushioned them as they bounced until they settled in one place. The airbags then deflated and retracted, allowing the rovers to deploy their solar panels ready for exploring.

When it came to Curiosity, however, airbags just weren't going to cut it. The rover is the size of a car and weighs about a ton (900 kg). Instead, it was lowered onto the surface using a futuristic-looking sky crane. The onboard Mars Descent Imager helped to measure the probe's distance from the ground, and at a height of 65 feet (20 m) rockets fired to maintain its altitude. Cords then deployed to deposit the rover on Mars. The rockets then fired again to remove the landing rig from the area, making sure that it didn't crash and contaminate the region the rover was sent to explore.

These successes are incredible feats of engineering, but Mars landings remain complex and risky. This was underlined in October 2016, when the Schiaparelli EDM lander inadvertently crashed into the Martian surface at more than 310 miles (500 km) per hour.

v The steps required to land the Curiosity rover on Mars during the so-called "seven minutes of terror."

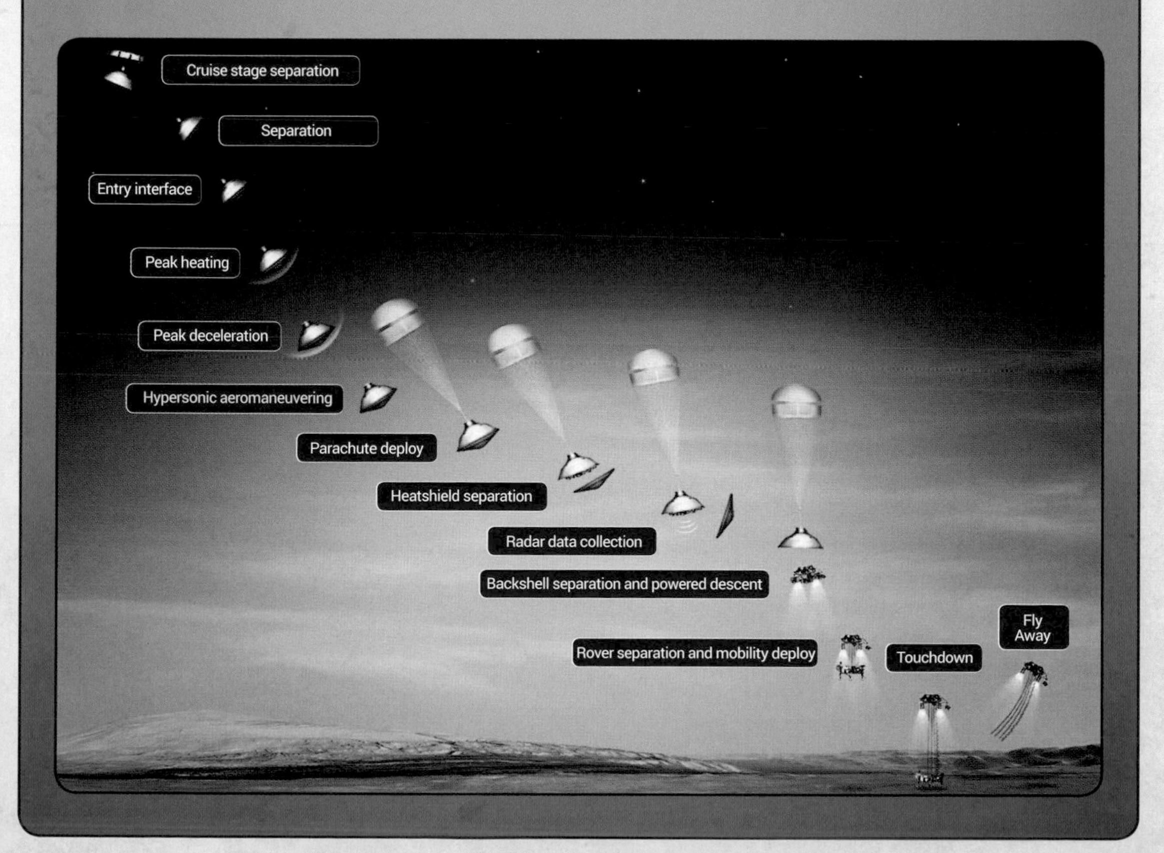

THE FIRST MARTIANS

AT THE END OF YOUR FIRST DAY, YOU WITNESS YOUR MAIDEN MARTIAN SUNSET. IT'S LIKE NOTHING YOU'VE EVER SEEN BEFORE—THE DAYTIME SKY IS RED, BUT IT TURNS BLUE AS THE SUN NEARS THE GROUND.

That's thanks to large quantities of dust present in the Martian atmosphere, which scatters the sunlight to create blue hues.

Don't expect sunset to be quick; even after the Sun has disappeared, the dust is still bouncing sunlight around the sky, meaning it takes an espeically long time to get dark on Mars. There is no rain on Mars to help wash dust out of the atmosphere, and the dust is not always helpful. Fierce dust storms the size of continents can engulf large swathes of the planet (see box, p. 162).

Unfortunately, time on the surface will be limited for the first travelers to Mars due to radiation dangers. Without a magnetic field or a thick atmosphere, the planet gives scant shelter from the ravages of high-energy particles from the Sun and exploding stars in the wider galaxy. The dose is the same as having a full-body CT scan every five or six days. For that reason, intrepid Mars travelers may well seek shelter underground. Photographs from orbiting space probes have already confirmed the presence of subterranean lava tubes carved out long ago by the flow of molten rock. Setting up house down there would provide a natural radiation shield. The tubes close to the Pavonis Mons volcano seem a particularly good bet.

↖ An artist's impression of an astronaut working on Mars.

> Sunset on Mars as seen by NASA's Spirit rover in 2005.

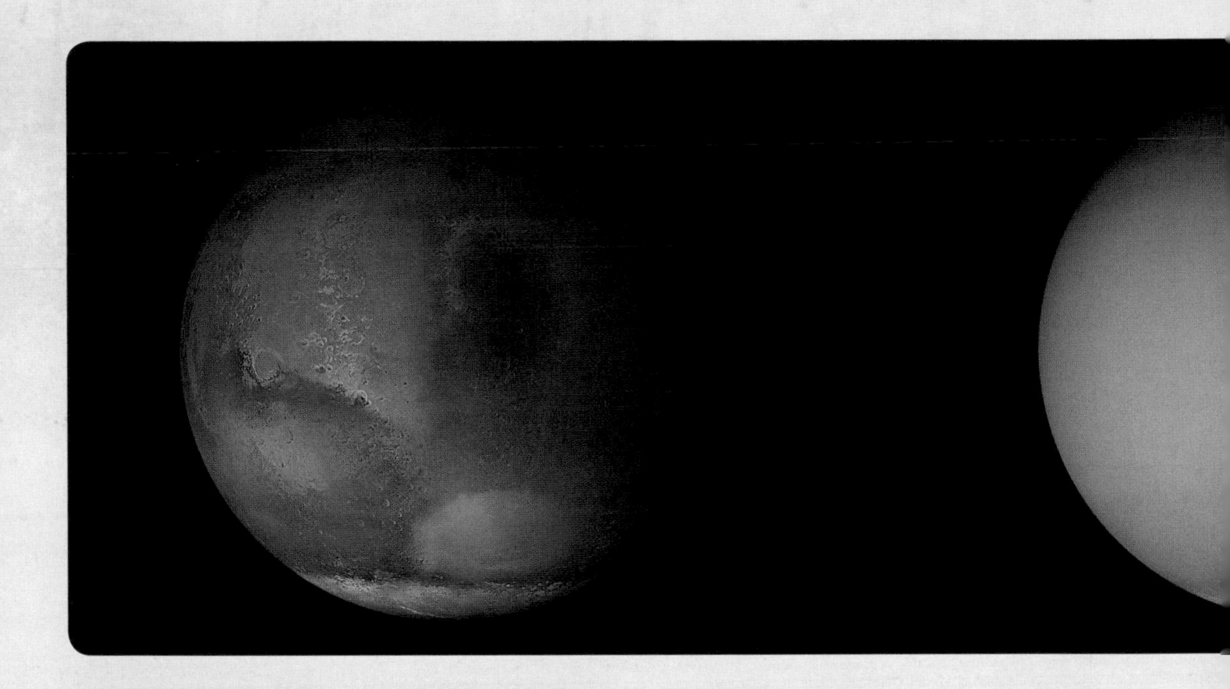

MARTIAN DUST STORMS

When NASA's Mariner 9 probe reached Mars in November 1971, mission scientists were eager to use its cameras to snap closeup photos of the planet's treasures. Unfortunately, the Red Planet was being camera-shy. Two months earlier, Mars had kicked up a mighty dust storm, and clouds of fine particles blanketed most of the planet and obscured the view. Only the tops of Mars's tallest volcanoes could be seen poking through. By January 1972, the storm had subsided, but it was a stark warning of the dangers of Martian dust.

Decades later, the Spirit and Opportunity rovers had to hunker down and go into hibernation in 2007 as regional dust storms raged across Mars. At one point, less than 1 percent of the usual amount of direct sunlight was reaching the Martian surface, seriously affecting the rovers' ability to draw power from their solar panels. Eventually those storms passed, too.

It is already clear that such dust storms would pose a threat to any future human missions to Mars. Global dust storms engulf the planet every three years. Not only would the dust coat solar panels, reducing their efficiency, it would also get into mechanical equipment and cause damage to gears and other moving parts.

Wearing your space suit at all times will be a must. The thin Martian atmosphere means that the air pressure on the surface is the same as that 15 miles (25 km) up in Earth's atmosphere—almost three times higher than the summit of Mount Everest. The lack of pressure results in the boiling point of water dropping to just 50 degrees Fahrenheit (10°C). Without a pressurized spacesuit, your body would be warm enough to boil the saliva on your tongue.

A lot of your time will be spent trying to cultivate plants in order to supply you and the crew with fresh and nutritious food. Artificial lighting may well be needed, because the sunlight is only 40 percent as bright on Mars as on Earth, due to its farther distance from the Sun. Studies are already being done to identify which crops will probably be best suited to the Martian environment. Soil was taken from a volcano on Hawaii to stand in for Martian dirt. Radishes, peas, tomatoes, cress, and arugula all seemed to thrive and did not absorb dangerous toxins during the experiment.

∧ These images from Mars Global Surveyor (MGS) show how dust storms engulfed the planet in 2001.

↗ Water ice in a crater at the Martian North Pole.

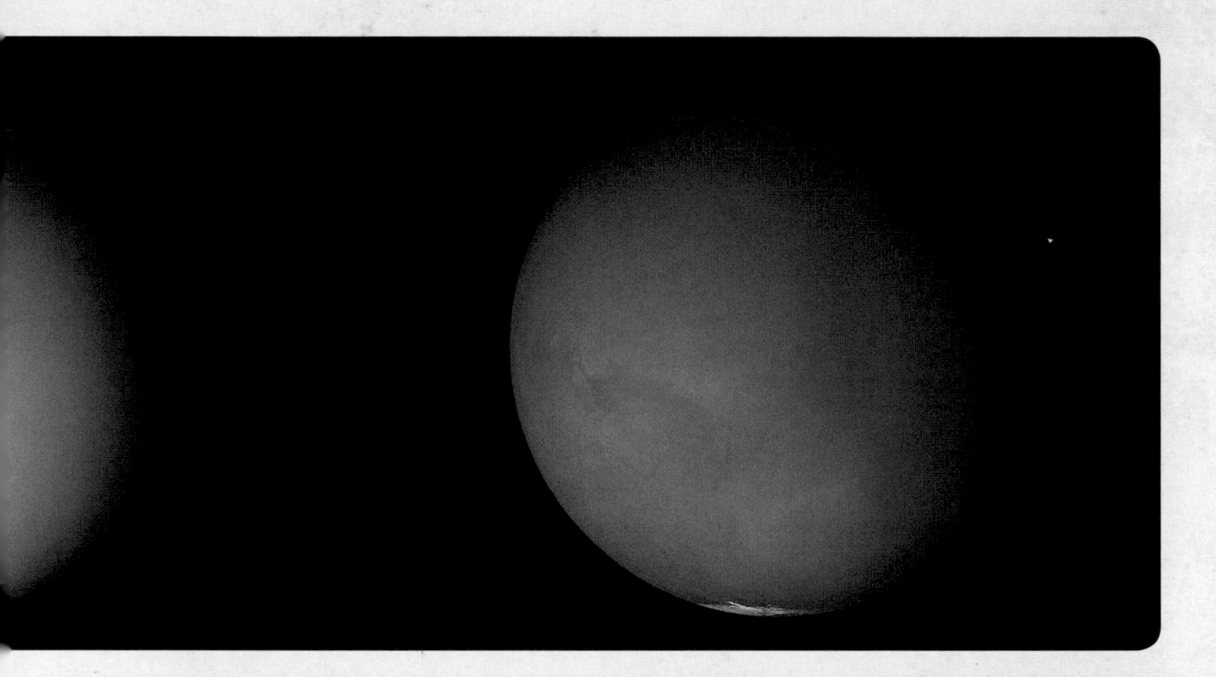

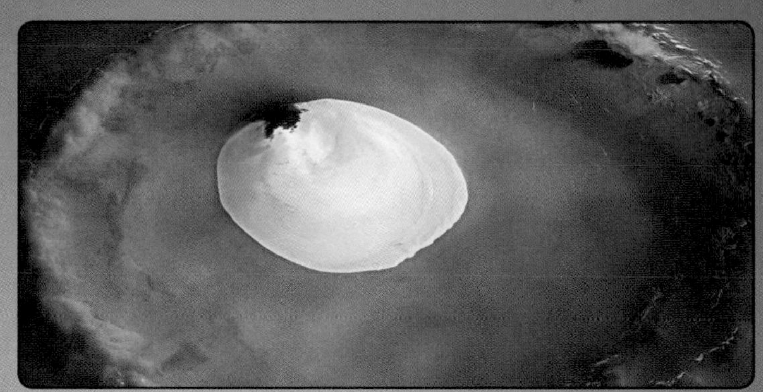

More than 300 million miles (500 million km) is a long way to drag everything you'll need to survive on the Martian surface. If you can use what's there, your mission will be lighter, less expensive, and have a greater chance of success. Weight is one of the biggest limiting factors for a successful landing on Mars (see p. 159).

Two of the key resources needed for human survival are water and oxygen. Fortunately, Mars has both in abundance—you just have to know how to get at them. We already know that the Red Planet has vast quantities of water ice. In 2016, NASA's *Mars Reconnaissance Orbiter* found enough subsurface ice in the planet's Utopia Planitia region to fill Lake Superior—the largest of the Great Lakes. Melt the ice and you have liquid water, but you're also one step closer to having oxygen. Chemically, water is H_2O: two atoms of hydrogen bonded to one atom of oxygen. A process called electrolysis can break those bonds by passing an electric current through the water. Oxygen could also be extracted from Mars's carbon dioxide (CO_2) atmosphere. MOXIE (Mars OXygen In Situ Resource Utilization Experiment) will fly on the planned Mars 2020 rover to test this idea.

LIFE ON THE SURFACE

MARS IS NOT SHORT OF SPECTACULAR SIGHTS FOR THE INTREPID VISITOR. THE PLANET'S STANDOUT GEOLOGICAL FEATURE IS OLYMPUS MONS—THE SOLAR SYSTEM'S TALLEST VOLCANO AND ITS SECOND HIGHEST PEAK.

Olympus Mons towers a colossal 69,839 feet (21,287 m) above the Martian surface, making it nearly two-and-a-half times higher than Mount Everest here on Earth. At 373 miles (600 km) across, it is about the width of France, and it covers a total area approximately equal to that of Italy.

Despite its impressive dimensions, the slopes of Olympus Mons are very shallow, with an average incline of just 5 degrees. That means you wouldn't be able to see the top from the bottom, as the volcano's peak would be hidden over your horizon. It is unclear if the volcano is still active, with some suggestion of lava flows a few million years ago.

> A composite orbiter image of Olympus Mons on Mars, the tallest known volcano in the Solar System.

MARTIAN ASTRONOMY

The Martian sky has many similarities to Earth's, but with some notable exceptions. You'd see the same stars in their familiar constellations, but they're joined by our planet appearing as a blue visitor in the morning or evening sky. Our Moon is big enough to be seen alongside Earth, but it regularly disappears behind or in front of our blue-marble planet as it orbits us. One of the rarest sights would be the transit of Earth, when our planet appears to pass directly in front of the Sun. The next one will take place on November 10, 2084.

Mars's own moons—Phobos and Deimos—whizz across the sky, too. Phobos, the largest of the two, orbits the Red Planet in just eight hours, which means it rises and sets twice every day. It has phases, but because of its swift passage, they change every hour. Although Phobos occasionally eclipses the Sun, it isn't big enough to block out the Sun entirely.

Mars is closer to Jupiter and Saturn, so both planets will appear larger through telescopes than when on Earth. You won't need as big a telescope to spy Jupiter's Great Red Spot or Saturn's rings.

v Earth and the Moon as seen from Mars by the Mars Reconnaissance Orbiter.

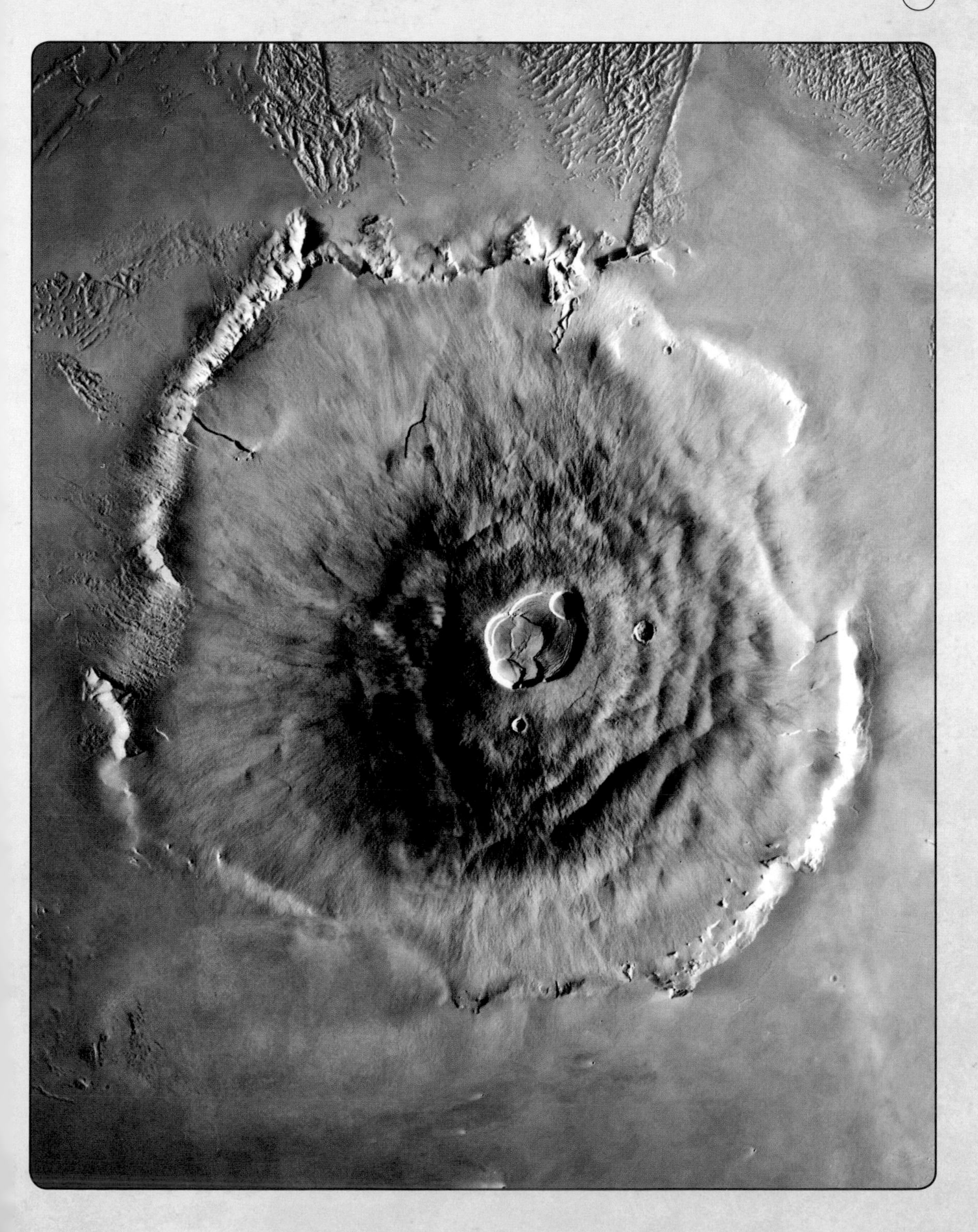

To the southeast of Olympus Mons, on the other side of three more volcanoes known as the Tharsis Montes, lie the Valles Marineris (The Mariner Valleys). Ancient fissures in the Martian surface, at more than 2,485 miles (4,000 km) long, they stretch almost one-quarter of the way around the Martian equator, from the Noctis Labyrinthus in the west to Chryse Planitia in the east. They're 124 miles (200 km) wide and 4⅓ miles (7 km) deep, easily eclipsing the majesty of the Grand Canyon in Arizona (which is only 18 miles/30 km wide and 1¼ miles/2 km deep).

Cross the Martian meridian into its eastern hemisphere and you'll encounter the Hellas Basin, one of the mightiest impact scars in the entire Solar System. It is believed to be the result of a calamitous collision around four billion years ago. The rim sits 29,528 feet (9,000 m) above the basin floor. Place Mount Everest in the crater and its peak wouldn't poke out above the edge. Such a low elevation leads to a significantly increased atmospheric pressure that's double the Martian average. Not only would that make it more suitable for human habitation, it is one of the few places on Mars where liquid water might be able to exist naturally.

The frigid temperature on the Red Planet means that there are many glaciers to explore. These slow-moving ice rivers are mostly found in two bands, each halfway between the Martian equator and the poles. An estimated 5,300 billion cubic feet (150 billion cu. m) of water is locked up in Mars's glaciers. The Martian ice represents an unprecedented spectacle of frozen carbon dioxide (dry ice). The northern cap is particularly spectacular, with the 310-mile (500-km)-long Chasma Boreale cutting it almost in half. It might even be possible to partake in some traditional winter sports on some parts of the Red Planet (see box, below).

< Global mosaic of 102 images of Mars, showing the Valles Marineris cutting across the Martian equator.

Once humans permanently settle on Mars, attention might well turn to fun and leisure time. Just as on Earth, sport may become a big part of the recreational calendar. One day there could be a Martian Olympics, with the records for jumping and throwing events much higher due to Mars's lower gravity. Some Olympic sports, such as golf, would need altering—the holes would be much longer and it would take an impractical amount of time to play all eighteen. A shortened form, such as pitch and putt, might have to do.

The slopes of the Martian ice caps may offer the best skiing on the planet. However, because of the weaker gravity, you'd have to start three times higher than on Earth to achieve the same slalom speed. You'd also have to ski in a space suit with an oxygen supply. The Bagnold Dune Field explored by NASA's Curiosity rover would make an excellent location for dune buggying and sand boarding. Mars represents a paradise for climbers, too. Most Martian rock is volcanic basalt, complete with vesicles—gaps created by bubbles in the original lava that make ideal finger holes. Equally, the calderas (collapsed craters) at the top of Olympus Mons would be the perfect spot for abseiling.

> The Bagnold Dunes, as seen by NASA's Curiosity rover during its 1,174th Martian day,

TERRAFORMING

THERE'S GROWING EVIDENCE THAT MARS WAS ONCE A MUCH WARMER AND WETTER PLANET. PERHAPS ONE DAY WE'LL MAKE IT THAT WAY AGAIN.

As it stands, however, Mars is an extremely hostile place for humans. No oxygen to breathe, in a carbon-dioxide atmosphere incapable of protecting you from the ravages of radiation. Nor can it provide sufficient atmospheric pressure to keep water in its all-important liquid form.

To put it bluntly, Mars is what a property developer might call a "fixer-upper." This once lush world has turned cold, rusty, and dusty, and is in much need of renovation to make it fit for habitation. Space scientists call this wholesale cosmic landscaping "terraforming."

Mars certainly has bags of potential. We know from our current climate issues on Earth that carbon dioxide is a potent greenhouse gas. It lets sunlight through, but acts as a blanket trapping that heat. Release some of the CO_2 locked up in Mars's polar caps, and the Martian temperature would rise slightly. In turn, that would release more CO_2, making the planet even warmer. Releasing other greenhouse gases, such as methane and ammonia, might work, too. It's a runaway process. Eventually, the atmospheric pressure and ambient temperature will be able to support liquid water. Rain will fall, and rivers will meander into oceans. Once water exists on the surface, we could introduce plants that would begin to photosynthesize, drawing in carbon dioxide and exchanging it for oxygen. Certain types of microbes and bacteria would do a similar job. We could end up with a blue–green planet not too dissimilar in appearance to our own.

THE MARS TRILOGY

The arena of science fiction offers the freedom to explore ideas that are more advanced than our technological capabilities in the real world. When it comes to terraforming Mars, the stories of Kim Stanley Robinson lead the way.

In 1993, he published *Red Mars*, the first in a trilogy of books about the planet. This was followed in 1994 by *Green Mars* and in 1996 by *Blue Mars*. The action starts in 2026 with the first human colony arriving, a crew of 100. As the trilogy develops there are debates about the morals and merits of terraforming Mars. Saxifrage "Sax" Russell believes it to be crucial, but Ann Clayborne argues against it on the grounds that we're arrogant to think we can do what we want to other worlds.

Terraforming does indeed go ahead, first by digging boreholes to release heat trapped underground, then by detonating nuclear devices to melt the Martian permafrost. By the time of *Green Mars*, set 50 years later, plants have started to grow. Orbital mirrors are installed to raise temperatures farther. By *Blue Mars*, liquid water can freely exist on the Martian surface and humans have started terraforming other worlds, including Mercury and Venus.

> What Mars may look like one day if we can successfully terraform it.

But it won't be happening right away. Just as those who started building the grandest cathedrals knew they wouldn't see them completed in their lifetimes, so terraforming Mars will be a multigenerational endeavor. It could take thousands of years to get to the stage where we no longer call Mars the Red Planet.

At present we do not have advanced enough technology to undertake such a bold move. That has not stopped science-fiction authors from filling in the gaps ahead of time, however. They've often explored the ethical issues with terraforming alongside the practical concerns (see box, p. 168). Scientists, too, have sketched out ways in which to do it. One of the leading ideas is to fashion a giant mirror to focus the Sun's rays down onto the polar caps (see box, right).

Yet none of these improvements will last unless we can generate the protective bubble of a magnetic field around Mars. Otherwise the solar wind will strip away all of our hard work over time. At present, there is no consensus about how this crucial step might be achieved.

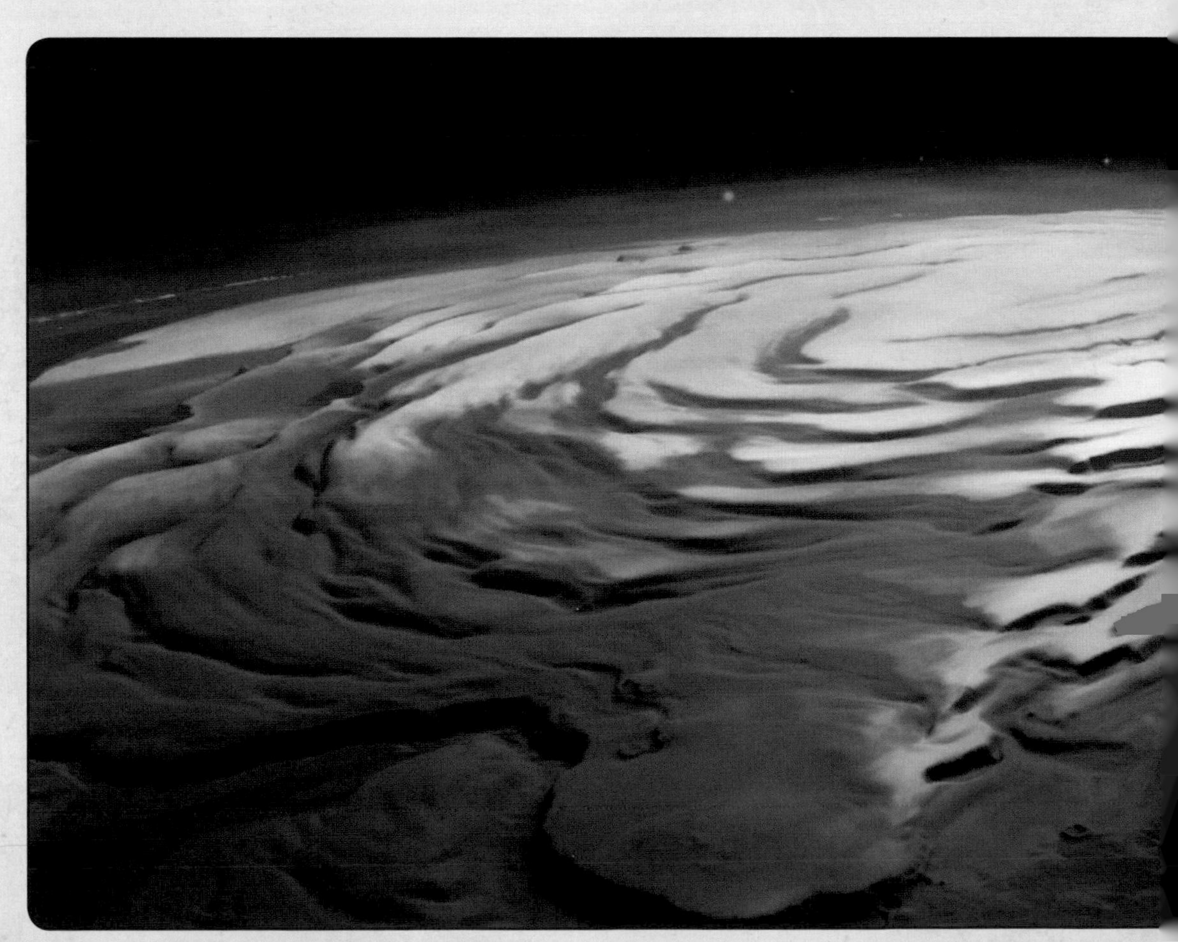

v The northern ice cap of Mars with the Chasma Boreale cutting through it, as seen by Mars Express.

< A mirror could focus the Sun's rays onto the Martian ice caps to raise the planet's temperature.

Reengineering a whole planet for our own purposes is no mean feat. The necessary technology is a long way off, but that hasn't stopped scientists from exploring how it might be done. A leading idea is to place a giant mirror close to Mars to focus the Sun's rays onto the planet's southern ice cap. A temperature rise of just five degrees Celsius would liberate a considerable amount of carbon dioxide into the Martian atmosphere. After several hundred years, it might be possible to raise Mars's atmospheric pressure to 500 millibars—half that found on Earth.

Studies suggest the mirror would have to be at least 62 miles (100 km) across, meaning it might weigh several hundred thousand tons. That's far too heavy to be launched from Earth. It would have to be constructed closer to Mars, either from material in the nearby asteroid belt or from Mars's moons Phobos and Deimos. It wouldn't necessarily have to be parked in orbit around the Red Planet. It could sit farther out, hovering in a spot where the pressure of sunlight and the gravity of Mars balance. Such devices are called statites—a portmanteau of "static satellite."

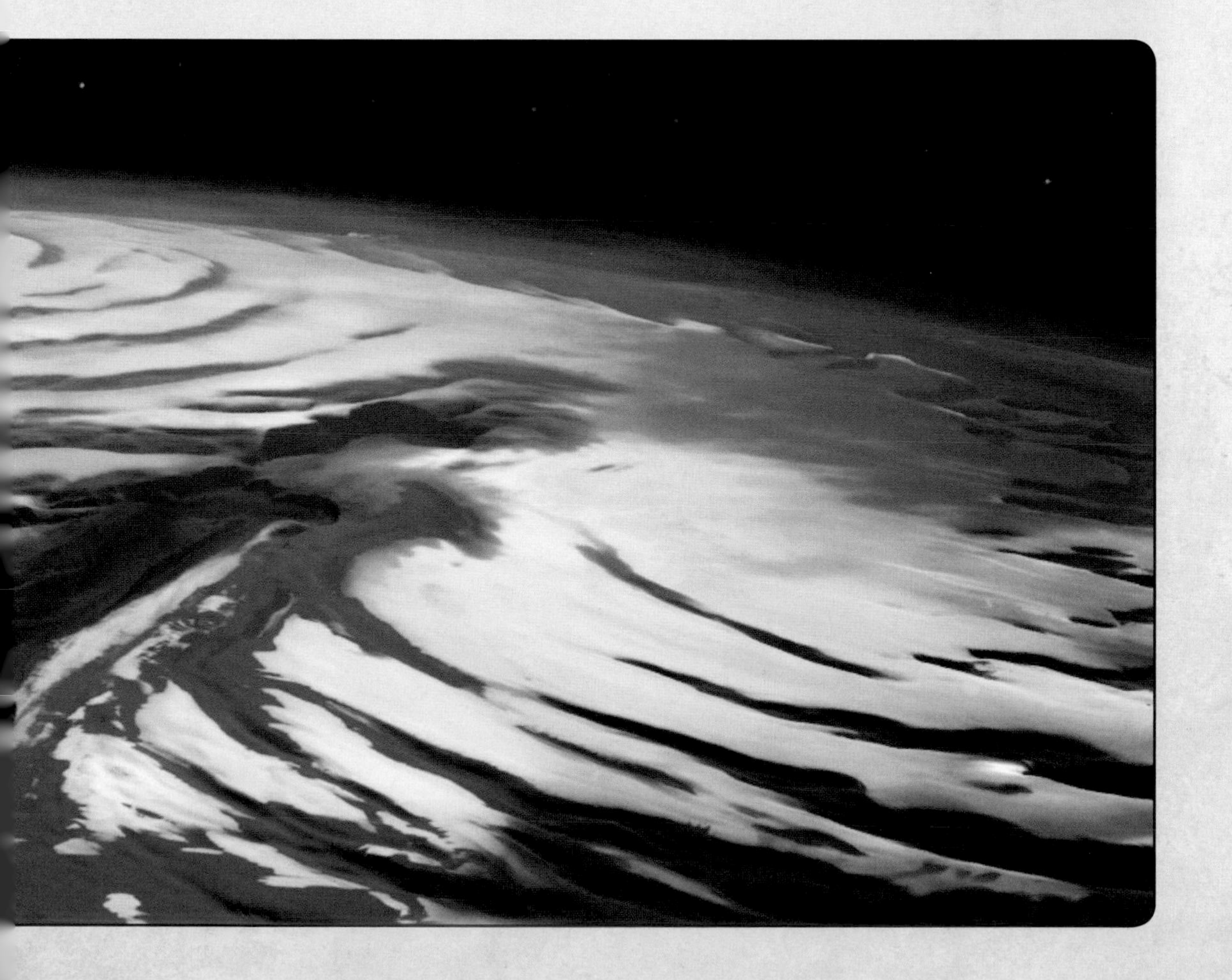

DEALING WITH A THREAT

OUR PLANET IS NOT ISOLATED IN SPACE. EARTH IS BOMBARDED BY ABOUT 110 TONS (100 TONNES) OF SPACE DUST EVERY SINGLE DAY.

About once a year, an asteroid the size of a car hits our atmosphere, flashing across the sky as a spectacular fireball. Luckily, it disintegrates before reaching the ground. Yet we haven't always been so fortunate.

Every couple of thousand years a monster the size of a football field strikes and causes extensive local damage. In 1908, a fireball exploded above the Tunguska River in Siberia and flattened 772 square miles (2,000 sq km) of forest. Had it struck just a few hours later, it would have devastated densely populated parts of Western Europe.

Then there are the behemoths that can threaten mass extinctions. Most notably, sixty-six million years ago, a 6-mile (10-km)-wide asteroid—the size of a small city—careered into the Mexican coast. Tsunamis raged across the sea, as dust and debris engulfed the sky. Our planet descended into a nuclear winter, with scant sunlight available even in the middle of the day. Starved of light, many plants died out. They were swiftly followed by the herbivores and then the carnivores. Within a century, 70 percent of all land species were wiped out. It was even worse in the oceans, where nine of out every ten species perished.

> An asteroid impact 66 million years ago has been implicated in the demise of the dinosaurs.

↗ A map of the world showing more recent impacts from space.

BOLIDE EVENTS 1994–2013

(SMALL ASTEROIDS THAT DISINTEGRATED IN EARTH'S ATMOSPHERE)

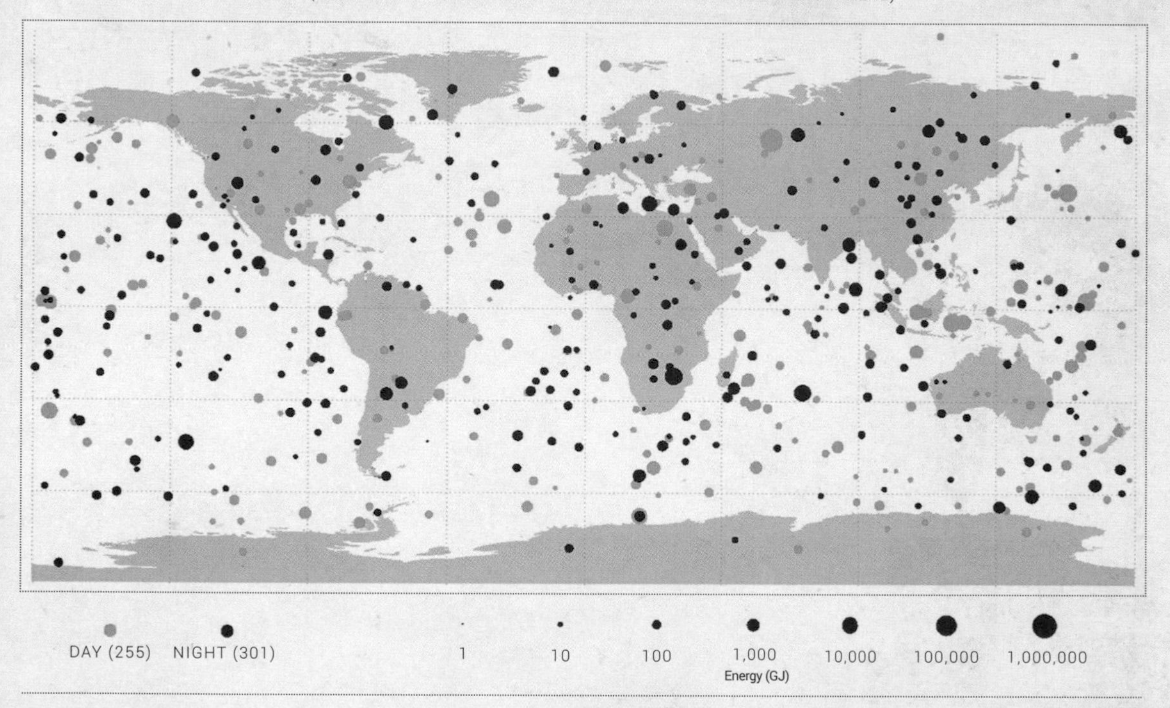

We obviously don't want to succumb to the same fate, and we're doing all we can to watch the Solar System for any potential threats (see box, p. 175). Yet that danger, along with the promise of mining riches (see p. 176), adds an inescapable lure to the human exploration of asteroids. It's no coincidence that space agencies around the world have discussed sending astronauts to asteroids as a stepping-stone on the road to Mars. Not only does an asteroid's small mass mean it is easier get on and off, but selling the cost of such missions to taxpayers is easier

HOW TO STOP AN ASTEROID COLLISION

Contrary to the ideas of Hollywood scriptwriters, blowing up an incoming asteroid is not the way to go. Doing so would just shatter it into smaller pieces still tearing through space toward us. The best option is to keep it in one piece, but to alter its path so that it's no longer a threat.

The earlier we spot the asteroid, the easier this is. With many years' notice we'd only have to change its speed by an inch or so per second. NASA is currently looking into two promising ways to achieve this. The first is a so-called kinetic impactor, shunting the asteroid by firing something into it to slow it down. The second tactic employs a "gravity tractor," using the gravitational pull of a space probe placed next to the asteroid to tug it off course. Future human missions to asteroids could help astronomers assess the composition of asteroids and investigate which of these methods would most probably succeed.

There have been other, more outlandish ideas suggested. They include exploding dye on one side of the asteroid to change the way it reflects sunlight, and attaching a tether to pull it away from the danger zone.

if you justify it as a way of working out how to deal with any potential danger. The idea of grabbing an asteroid and towing it closer to Earth for human exploration has also been mooted, with the added benefit of testing out robotic alternatives (see box).

Any astronauts would have to tether themselves to the asteroid at all times. This is because even the biggest asteroid in the Solar System—Ceres—has a surface gravity thirty-six times weaker than we experience on Earth. That means a 154-pound (70-kg) astronaut would weigh the same on Ceres as a 4½-pound (2-kg) bag of potatoes does on Earth. You might just be able to walk on Ceres, however, on smaller asteroids you'd run the real risk of breaking free of its gravity and drifting off into space.

SPOTTING A KILLER ASTEROID

< NASA's NEOWISE project is helping us track down large asteroids that might pose a threat to Earth.

We have one massive advantage over the dinosaurs: telescopes. NASA's Near Earth Object (NEO) program has already located more than 90 percent of all asteroids larger than 3,300 feet (1 km) in diameter. Once an asteroid is discovered, its path for the next one hundred years is calculated, to see if it might pose a threat to Earth. That threat is assessed on the Palermo Technical Impact Hazard Scale, and so far no significant known dangers have been identified. The asteroid 410777 (2009 FD) has the highest impact probability—1 in 714—and that's not until 2185. Additional observations will almost certainly see those odds lengthen.

However, that still leaves tens of thousands of Potentially Hazardous Asteroids (PHAs) yet to be discovered. NASA's new goal is to track down 90 percent of all asteroids more than 460 feet (140 m) across. Enter the Large Synoptic Survey Telescope (LSST). Due to start operating high in the Chilean desert in 2022, it will be able to meet that goal over its 10 years of intended operation. It will also be able to spot dangerous asteroids as small as 148 feet (45 m) across, with one to three months' notice. That's thanks in no small part to a 3.2-billion-pixel digital camera—the largest ever constructed.

LIFE IS HARD FOR BELTERS. THEY'RE FORCED TO WORK LONG, UNFORGIVING HOURS AT THE SEAM, SUPPLYING EARTH AND MARS WITH THE NATURAL RESOURCES REQUIRED TO SUSTAIN HUMAN CIVILIZATION ACROSS TWO PLANETS.

People on both worlds see them as the lowest of the low—the downtrodden working class of the Solar System.

Being born in the asteroid belt means that these people are taller and thinner due to the decreased gravity. Childbirth is particularly arduous due to the mothers' weakened muscles. Spies are tortured by bringing them down to Earth, where the much stronger gravity cripples their fragile bodies. This so-far-imagined reality takes place in *The Expanse*, a recent sci-fi television series about life in space set two hundred years in the future. Earth and Mars are heading for war, and the Belters are stuck in the middle.

Such a scenario is not as farfetched as it might first appear. The lure of asteroids is inevitable. In 2015, the space rock 2011 UW158 flew by Earth at a distance six times farther than the Moon. Estimates suggest it could

> Artist's concept of NASA's OSIRIS-REx mission collecting a sample from the asteroid (101955) Bennu.

PLANETARY RESOURCES, INC.

Asteroids could be big business, and companies already exist that are looking for a slice of the action. One of the early front-runners is Planetary Resources, Inc., a company whose employees include many former NASA engineers as well as advisors, including movie director and explorer James Cameron. Early investors include Google founder Larry Page and Virgin Group CEO Richard Branson.

In January 2018, Planetary Resources launched the Arkyd-6 satellite into orbit on top of an Indian rocket. The size of an ink-jet printer, this satellite is equipped with infrared cameras and is a test of the technology required to survey the near-Earth asteroid population for suitable mining targets. It is looking for water, in particular, because water's chemical structure of two hydrogen atoms bonded to a single oxygen atom means it can yield both hydrogen fuel and liquid oxygen propellant, along with breathable air—all crucial ingredients if humans want to colonize the Solar System.

The company plan to launch multiple spacecraft to several target asteroids in 2020. On arrival, they will collect data and test material samples. What they find will be fed back into the design of the first commercial mine in space.

EARTH RESCUE MISSION

contain $5 trillion worth of precious metals, including platinum, gold, and silver. Our robotic missions to the asteroids Ceres, Vesta, and Itokawa (see box, below) have revealed plenty of water ice, too, which can be converted into both oxygen and fuel. The NASA probe OSIRIS-REx was launched in September 2016 and is due to rendezvous with the asteroid (101955) Bennu in 2023, grabbing a sample and returning it to Earth. Asteroids will become the cosmic mines and gas stations of the future.

Commercial space mining is already starting. In 2017, Luxembourg became the first European country to pass a law related to the utilization of space assets. The new legal framework recognizes the right to extract space resources in compliance with existing international law, notably the 1967 Outer Space Treaty (see pp.112–15). Space mining companies already exist and have plans to test the technology required to select and harvest asteroids for their riches (see box, p. 176). Eventually humans may live and work in the Asteroid Belt, much as workers live on oil rigs out in the sea today.

There are significant issues, however, that need to be addressed. Flooding the market with space gold would make gold less scarce, increasing supply and decreasing demand. The price of gold would likely drop, possibly affecting the profitability of the whole endeavor. Regulation would have to be tight to stop the rise of monopolies. Moving asteroids around the Solar System to get at them more easily could also be risky. One wrong move and you could drag the asteroid into an orbit that would see it collide with Earth. However, it is hard to see a future without some form of asteroid mining.

< An artist's impression of a spacecraft mining an asteroid near Earth.

HAYABUSA AND ITOKAWA

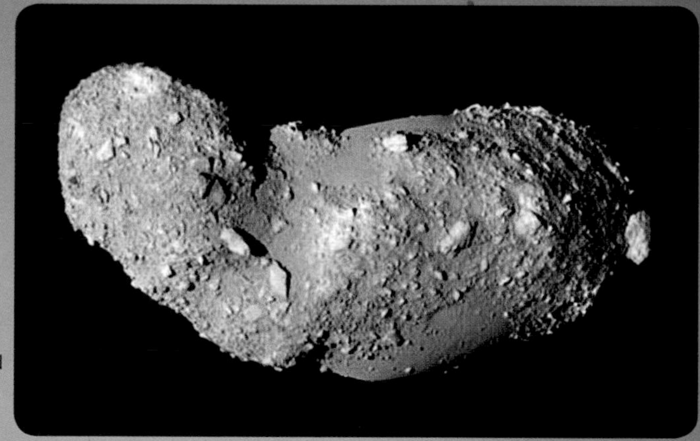

> The Japanese probe Hayabusa visited and took a sample of the asteroid Itokawa, shown here.

If asteroid mining sounds like science fiction, it isn't as futuristic as it first appears. We've already sent a probe to an asteroid, collected a sample, and returned it to Earth for analysis.

The Japanese Space Agency (JAXA) dispatched the Hayabusa probe to rendezvous with an asteroid called 25143 Itokawa in 2005. Although not everything went to plan, it did manage to grab some material from the surface and launch back toward Earth. On June 13, 2010, it crashed back through the atmosphere and landed in the Australian Outback. On the way down, it experienced a deceleration force of around 25 g, and the heat generated through friction with the air was thirty times the level experienced by the returning Apollo astronauts. Further analysis revealed that a haul of 1,500 dust grains from Itokawa had survived, each around 0.01 millimeter across.

Despite this success, the mission was a reminder of how difficult deep space maneuvers remain. Just a few months after launch, the craft was crippled by a huge solar flare that affected its solar panels and engine power. A communications failure then led Minerva—a probe Hayabusa had carried to land on the asteroid—to miss its target entirely and irretrievably drift off into space.

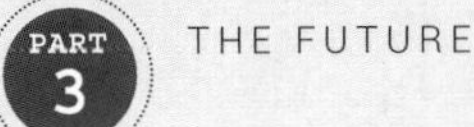

INTERSTELLAR TRAVEL

THE NAME "ASTRONAUT" IS A MISNOMER. IT MEANS "STAR SAILOR," BUT HUMANS HAVE SO FAR NEVER COME CLOSE TO MAKING OUR WORD FOR SPACE TRAVELERS TRUE TO ITS MEANING.

The farthest we've been is to the Moon, only 240,000 miles (385,000 km) away. By contrast, the nearest star to Earth after the Sun—Proxima Centauri—lies 4.2 light-years (or 24.7 trillion miles/40 trillion km) distant from us. Humans have traveled just 0.000001 percent of that distance.

With our current rocket technology it would take us tens of thousands of years to get there. In order to complete the trip in a human lifetime, we'd need to travel at a healthy fraction of the speed of light; 10 percent would see us arrive in forty-two years. However, propelling a spaceship to those kinds of speeds requires an amount of energy currently beyond our means. Accelerating a single ton to 10 percent the speed of light would take about 450 petajoules of energy. That's roughly the same as the annual energy consumption of a small country. A human mission would weigh quite a few tons. NASA's retired Space Shuttles weighed 83 tons (75 tonnes) when empty. We'll probably need more energy than the entire world currently uses in a year.

These significant hurdles haven't stopped scientists and science fiction writers from speculating how it might be done. Traditional chemical rockets just won't cut it. We could use the controlled detonations of nuclear devices instead, but current space law all but prevents such mechanisms from being tested. Better solutions would involve propulsion methods that pack a lot of punch for not a lot of fuel mass, keeping weight down.

Antimatter rockets

Every particle of ordinary matter has a mirror antimatter particle with the opposite electrical charge. The antimatter equivalent of the electron, for example, is the positron. When a particle meets its antiparticle, they annihilate each other, creating a burst of energy in accordance with Albert Einstein's equation $E = mc^2$. Antimatter rockets would use this energy to propel a spacecraft.

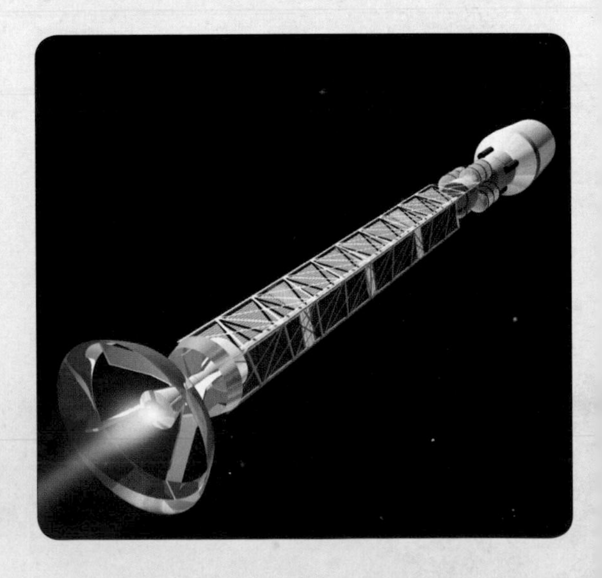

Fusion engines

Maybe we could copy the way the Sun makes energy. The nuclei of atoms fuse together deep inside our star, liberating vast quantities of energy as a result. In 2017, NASA gave start-up firm Princeton Satellite Systems a grant to pursue fusion rocket technology. However, the method would require a supply of helium-3, which is rare on Earth.

↲ An artist's rendition of an antimatter propulsion system.

v A concept for a fusion rocket that could be used to reach Mars more quickly.

BREAKTHROUGH STARSHOT

Sending a human crew to another Solar System is a long way off; it may not happen for centuries. Getting a space probe there, however, is a different story. The Breakthrough Starshot project is investigating just this possibility and aims to launch for the Proxima Centauri system within a generation.

With computer technology shrinking all the time, what used to take up an entire room can now fit onto a tiny microchip. Sails could be attached to nano-size spacecraft and laser beams fired from Earth to propel them into the void at speeds topping 93 million miles (150 million km) per hour. That's swift enough to get to Proxima Centauri in just thirty years. Pictures of its planets and other scientifically valuable data would take only 4.2 years to be beamed back at the speed of light.

Each "StarChip" could be produced for the same cost as an iPhone and many of them sent en masse to make sure at least some of them survive the treacherous journey; each probe could encounter about one thousand cosmic dust particles along the way. Scientists are currently working on some of the other outstanding challenges, including battery life, legal obstacles, and the necessary laser technology.

Ramjets

Rockets are heavy because they need to carry all the fuel they'll need for the whole journey. But what if you could fill up en route? That's the idea behind ramjets, spacecraft that scoop up hydrogen gas from interstellar clouds as they go. The hydrogen could then be used in a fusion engine to generate thrust.

↲ Sails and lasers could be used to propel micro-spacecraft to the next Solar System.

v A ramjet would collect gas from empty space to use as fuel.

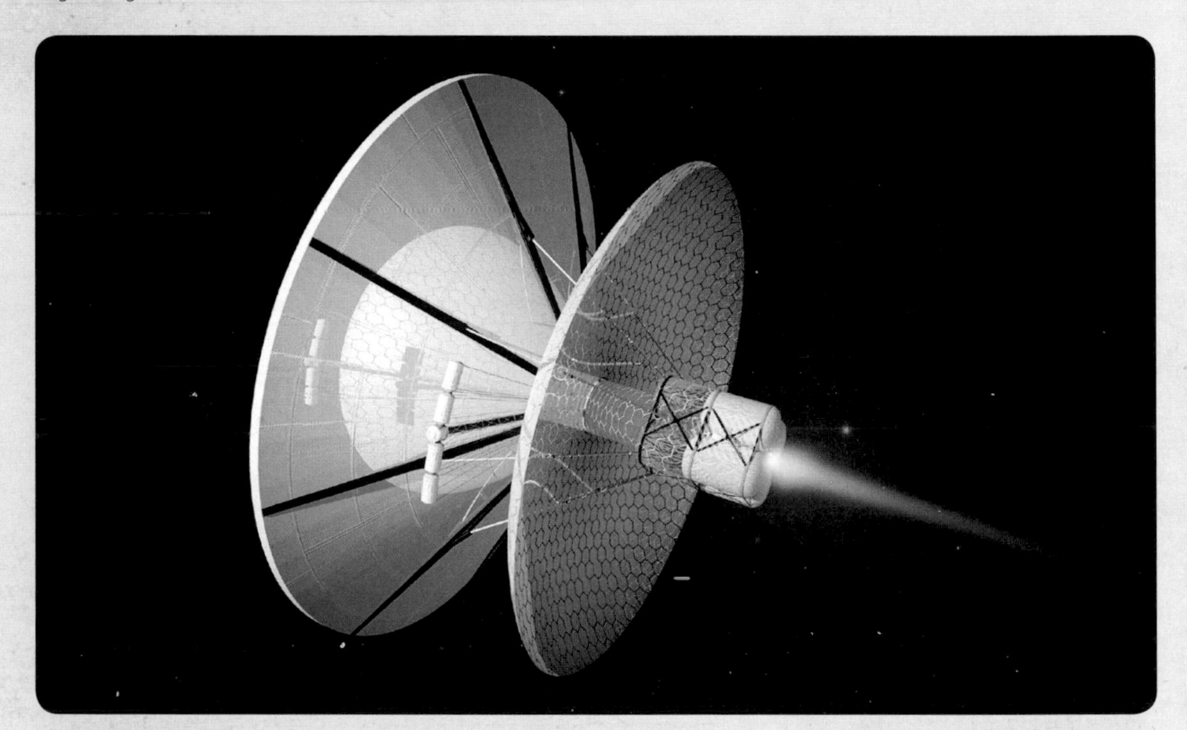

Most methods for traveling to other solar systems require spacecraft to attain a significant fraction of the speed of light to get there in a human lifetime. But what if speed wasn't an issue? Various science-fiction writers have turned to the idea of a "generation ship" to get around the problem. Here astronauts set off, live, reproduce, and die on board. Their children do the same, and then their children. The crew that eventually arrives at the other end hundreds of years later are the distant descendants of the initial astronauts.

Studies have even been done to look at how big the initial crew would need to be in order to ensure genetic diversity—so that you don't breed with someone too closely related to you. Estimates vary between 75 and 150. There is, however, a big ethical issue to consider. Think about the people living in the middle of the journey. They would have no memory of Earth and would never see the destination. They would effectively be prisoners on a spaceship, enslaved to reproduce. An alternative has been suggested in which frozen eggs and sperm are sent on the journey, with robots then fertilizing them thirty years before arrival. That, too, is riddled with ethical concerns.

GETTING THROUGH SPACE TAKES TIME. EVEN TRAVELING AT HALF THE SPEED OF LIGHT, IT WOULD TAKE MORE THAN EIGHT YEARS TO GET TO THE NEXT STAR AFTER THE SUN.

Yet a closer look at the theories of Albert Einstein reveals that your time as an astronaut on board a fast-moving ship would not be the same as for those of us who remain here on Earth.

Einstein published his special theory of relativity in 1905 and his general theory of relativity in 1915. He said that the three dimensions of space and one of time are wrapped up together into a four-dimensional fabric he called space-time.

GENNADY PADALKA—HUMANITY'S GREATEST TIME TRAVELER

Not only is time dilation an established scientific fact, time travelers already walk among us. Russian cosmonaut Gennady Padalka is the greatest time traveler in human history, a modern day Marty McFly. He holds the record for the greatest number of days spent orbiting Earth, a total of 879 days across several missions to Mir and the International Space Station.

Traveling through space(time) faster than us on the ground, less time passed for him than for us. Upon returning to Earth, he was 0.02 seconds younger than if he hadn't become an astronaut. Reaching the future faster than us, he time-traveled forward one-fiftieth of a second. That may not sound particularly impressive, but the difference was so meagre only because he was traveling at 17,400 miles (28,000 km) per hour relative to the ground. To extend the time difference out to days, weeks, months, and even years will require us to travel at considerably higher speeds.

One day in the distant future, we might dispatch a human crew at a considerable fraction of light speed. At about 87 percent of light speed, the returning crew will have seen thirty years go by, but we Earthlings will be sixty years older.

∧ Cosmonaut Gennady Padalka is the greatest time traveler in human history.

Two people moving through space-time at different speeds will disagree on how much time has passed. This effect, known as time dilation, has a negligible effect at small velocities but kicks in at significant fractions of light speed. If you travel to Proxima Centauri at 95 percent of the speed of light, you might expect the journey to take a little more than 4.2 years. And it's true, that's how long your journey would seem to people on Earth. Yet, thanks to time dilation, to you, the journey would take only 1.38 years.

You might at first think that this violates the sacred rule that nothing can travel through space faster than light. Indeed, by covering 4.2 light-years in just 1.38 years, it looks as if you have. In reality, however, another effect explained by Einstein kicks in: length contraction. As an astronaut traveling at high speed, you'd measure the distance to your destination as shorter than someone measuring it from Earth. At 95 percent of light speed, Proxima Centauri would be 1.31 light-years away to you; from your perspective the trip would be shorter. As far as you're concerned, you haven't broken any rules. That's why it's called relativity. Everyone's experience of space and time is relative and equally valid.

v Einstein pictured gravity as the result of massive objects warping the space around them.

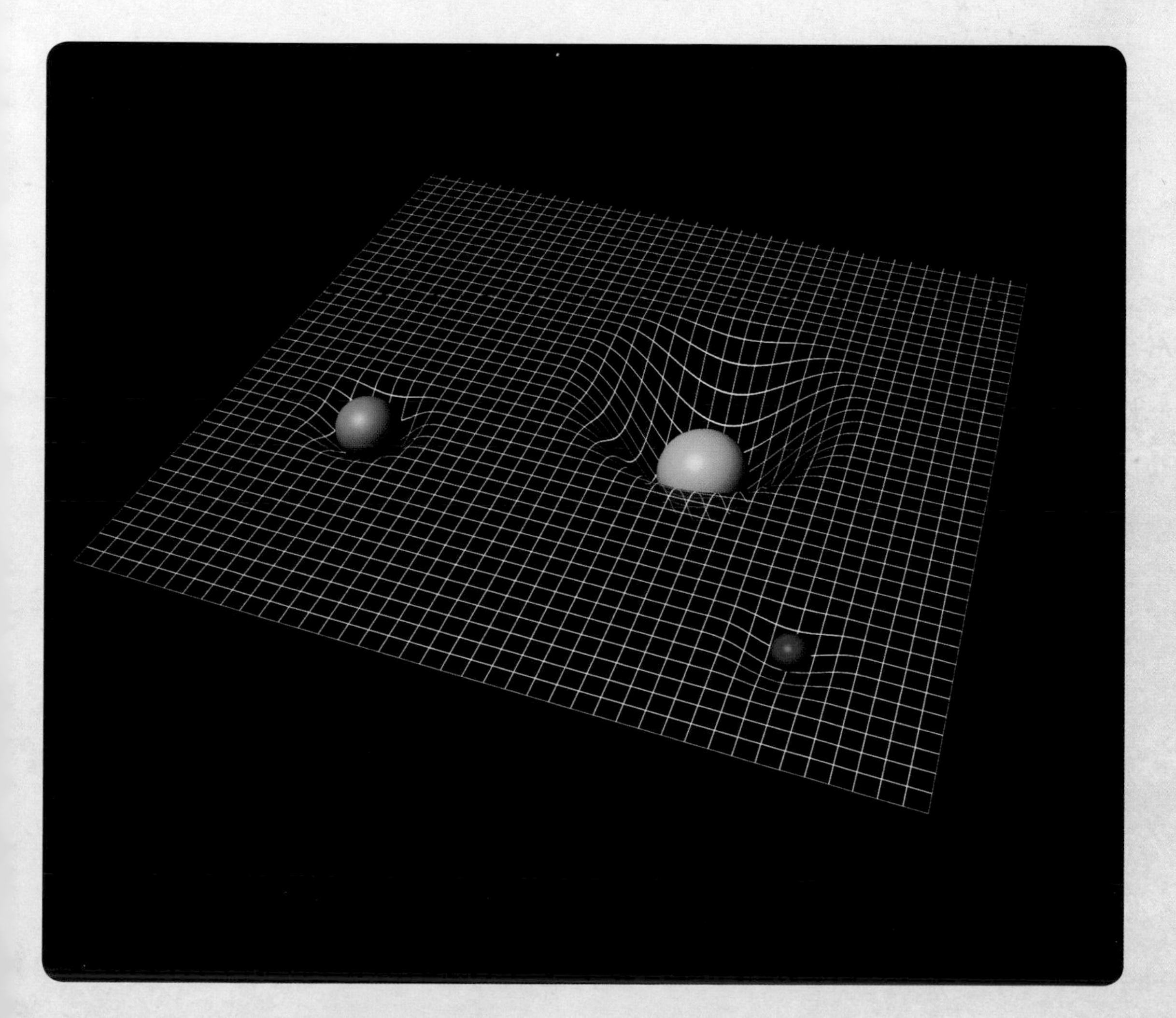

The upshot of all this is that the closer you can get to light speed, the less time the journey will take you. Yet a lot more time will always have passed back on Earth. So time dilation leads to the real possibility of time travel to the future. Imagine returning to Earth after a long, fast loop around space. You might have aged by only ten years, but on Earth a thousand years will have passed. You'd get to see the year 3000 without the need for a TARDIS or a DeLorean car. When it comes to time traveling to the future, a fast-moving spaceship is the only thing required. In fact, work in this area is already beginning to happen, and time travelers currently walk among us (see box, p.184).

> Shortcuts in space-time called wormholes could offer us a way to travel to the past.

Space is only travel-defyingly big if you take the conventional approach. Imagine a sheet of paper with Earth at one end and your destination at the other. Normally, you'd have to travel the full length of the paper to get there. In space, that might take thousands of years. But there is another way. If you fold the paper in half, then, suddenly, your destination is considerably closer. All you have to do is jump that short gap instead of traveling the long way round. Such a shortcut is known as an Einstein–Rosen bridge—or a "wormhole."

At the moment, wormholes remain entirely theoretical, but they do crop up in Albert Einstein's general theory of relativity, an idea that has so far passed every test thrown at it with flying colors. Some physicists even argue that you could use wormholes to travel to the past. However, if they do exist, they're probably only microscopic. We'd need a way of pumping them full of some kind of exotic energy if we want to make them big enough to be traversable by humans. But it's not impossible that a far more advanced civilization somewhere else in the universe has already worked out a way to rearrange the traditional astronomical architecture.

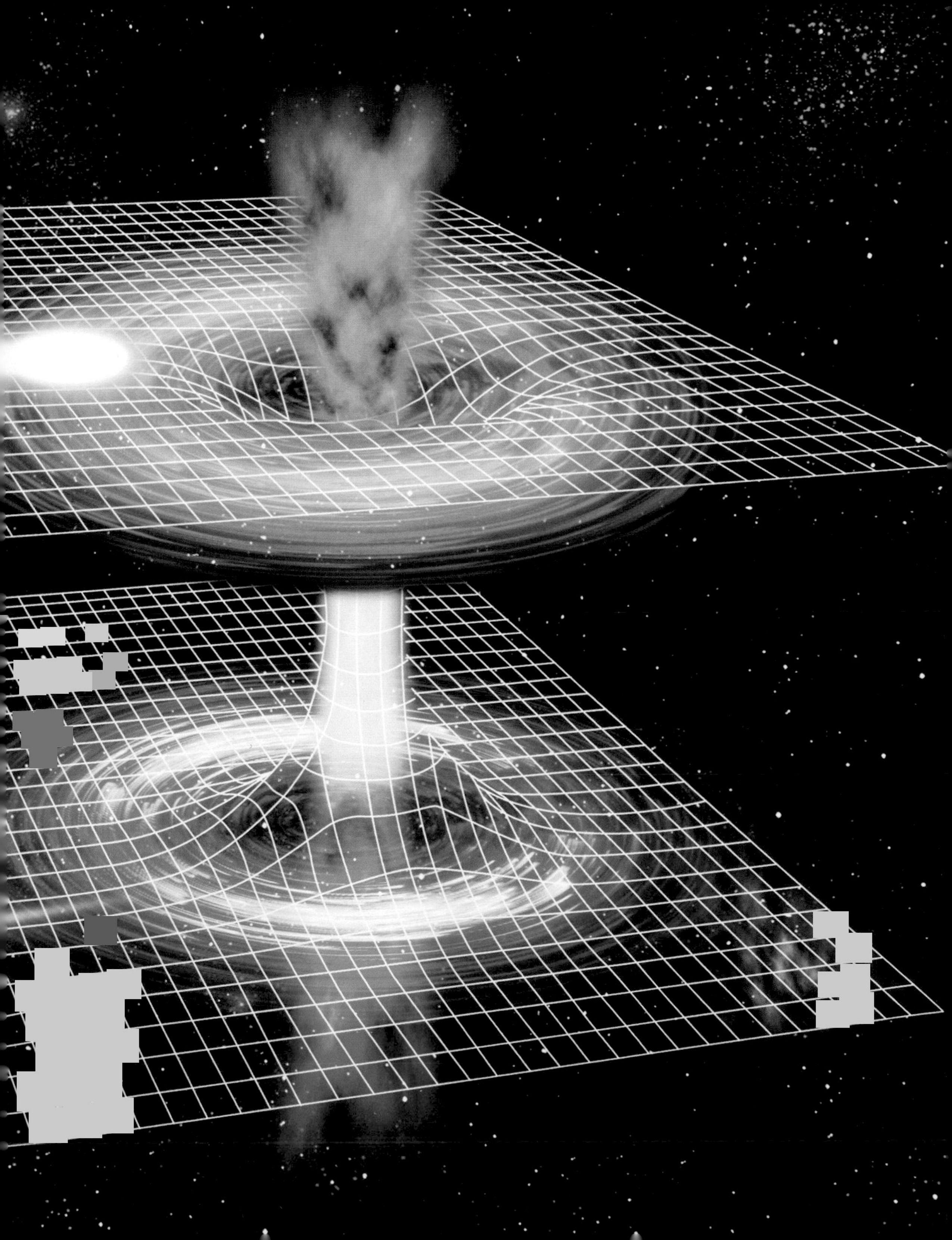

AS WE STAND AT THE BEGINNING OF A NEW CHAPTER IN THE HISTORY OF THE HUMAN RACE, IT IS IMPORTANT FOR US TO LOOK BACK OVER THE PREVIOUS ONES.

Without studying our past mistakes, we are doomed to repeat them. If we do things right, however, our future in space can benefit all of humanity.

The story of our journey out of Africa, across the world, and then beyond Earth's atmosphere is a messy one. Our history is littered with events of which we aren't particularly proud. The colonization of space offers us the chance to start fresh. To leave behind empires, wars, slavery, and environmental pollution in favor of a peaceful new frontier. The early pioneers of space law surely had this in mind when they drew up the 1967 UN Outer Space Treaty, just five years after the Cuban Missile Crisis brought the world to the brink of a nuclear war.

There's no way that it is going to be perfect. Progress is always a meandering path rather than a straight line. We could take a step backward before we take a leap forward. How do we control the way things are going? Our capitalist system will spread to space with all of its benefits and challenges. Some commentators fear that space is now becoming a playground for a small band of billionaire playboys. They argue that spacefaring is just another way for the rich to get richer, increasing the inequality gap between the haves and the have-nots. However, this doesn't have to be the case.

National governments are increasingly taking a back seat while private companies drive space exploration forward. It is easier to get shareholders to invest their money for profit than it is to convince voters of the need to invest tax dollars in space exploration instead of schools and hospitals. It remains to be seen if the state can effectively regulate this new arena. Too much regulation can kill innovation dead, yet too little will leave an opening for those hungry to exploit the resources of the Solar System solely for financial gain. The pace of change is quickening to such an extent that the traditional way governments enact legislation may have to be rethought. By the time you've drafted a bill, introduced it, debated it, and voted on it, there have been new developments that make your legislation obsolete. Companies such as SpaceX and Planetary Resources, Inc. may become the Google and Facebook of tomorrow—huge multinational corporations that individual governments struggle to rein in.

So what do we do? Our exploration of space is a relentless juggernaut that in all likelihood cannot be stopped. Adventure and discovery (and greed) are just too hardwired into the human psyche for us to resist the chance to reach outward into the cosmos. So, instead of decrying it, we should be discussing what we want to achieve and how best to do that without spreading the darker aspects of human nature farther into space.

The future is incredibly exciting, but the challenges ahead are great. We'll probably have a permanent human colony on Mars within a few centuries. We may even start terraforming it to make it fitter for habitation. Yet we

should be prepared for a time when this civilization will want to be independent. Just as the United States fought to break away from the rule of King George III of England, Mars may one day want to govern itself without rule from Earth. If we don't handle the situation carefully, we could end up with a war between planets, a continuation of our past mistakes. Interplanetary diplomacy could be a key skill in the centuries to come.

Looking even further ahead, we may want to leave the Solar System entirely and send the first people on an interstellar voyage to neighboring stars. One day, we could travel at a significant fraction of the speed of light to get there,

∧ Humans will probably explore Mars for the first time later this century.

leading to a level of time dilation that sees the astronauts return to a planet hundreds of years older than when they left. Imagine dropping a person from the seventeenth century into the middle of twenty-first-century Manhattan. It's hard to see how they would cope. Throughout these pages, we've seen how space travel affects our fragile biology, but our future beyond Earth is just as much about these psychological and societal issues. Only by conquering those can we build a future in space fit for us all.

INDEX

The publishers would like to thank the following sources for their kind permission to reproduce the pictures in this book.
Key: t = top, b = bottom, c = center, l = left, and r = right

Daein Ballard: 169

© Canadian Space Agency: 17t, 18-19

DARPA: 54

ESA: 48, 49, 111, 149t, 149b; /D. Baumbach: 35; / C.Carreau: 185; /S. Corvaja: 17c, 25, 36c, 38b; /DLR/FU Berlin: 157, 170-171; /David Ducros: 110b; /Foster & Partners: 145, 146-147; /GCTC: 34-35; /IBMP. 148

Getty Images: Daniel Berehulak: 124-125; /Corbis: 113c; /Mark Garlick: 172; /Alexander Legaree: 11; /Central Press: 81; /KrisCole: 189; /MPI: 49b; /Popperfoto: 8; /Roger Ressmeyer/Corbis/VCG: 98; /Science Photo Library: 134-135; /Sovfoto/UIG: 13, 104t; /Time Life Pictures/NASA: 70, 97r

ISAS: 179

Public domain: 14, 45b, 113t

Rex Features: 150-151

Shutterstock: 4-5; /Tommy Alven: 26b; /BeeBright: 142; /Edobric: 186-187; /Graphiteska: 68, 86

Science Photo Library: 27b, 29, 85; /Chris Butler: 108; /Christian Darkin: 132; /Novespace/CNES/DLR/ESA: 31; /Detlev Van Ravenswaay: 118l, 129, 178

Courtesy of SpaceX: 7t, 7c, 152-153, 154, 155, 158

NASA: 7r, 16b, 19, 20, 21, 22-23, 26-27, 32t, 32b, 36b, 37, 39, 40, 41, 42, 43, 44, 45t, 46, 47, 50t, 50b, 51, 52, 53t, 53b, 54t, 56, 57t, 57b, 58, 59, 60t, 61b, 62, 63t, 63b, 64, 65t, 65b, 66, 67, 68, 71, 72, 73t, 73b, 74, 75, 77, 80, 82, 84, 86t, 86b, 88, 88t, 88b, 90-91, 92, 93, 94, 96, 97l, 99, 100, 101, 102, 103, 105b, 106t, 106b, 109, 111, 114-115, 116, 117b, 119r, 120-121, 127, 128, 130, 131, 137, 138, 139, 141, 142, 145t, 159, 161, 164, 165, 182, 183, 184; /Jim Campbell/Aero-News Network: 30; /Carla Cioffi: 38t; /Tracy Caldwell Dyson: 95; /ESA: 107; /Goddard Space Flight Center: 76, 177; /W.Ingalls: 17t; /iGoal Animation: 110t; /JPL: 174-175; /JPL-Caltech: 144; /JPL/MSSS: 162-163, 167; /JPL/Texas A&M/Cornell: 160; /Timothy Kopra: 83t; /MSFC: 180; /Planetary Science: 173; /SDO: 79; /University of Washington, MSNW: 181; /USGS: 166; /Mark Widick: 21t

Virgin Galactic: 126-127

Rigel Woida: 170t

Every effort has been made to acknowledge correctly and contact the source and/or copyright holder of each picture and Carlton Publishing Group apologizes for any unintentional errors or omissions that will be corrected in future editions of this book.